M52

Water Conservation Programs – A Planning Manual

Second Edition

American Water Works Association

Manual of Water Supply Practices—M52, Second Edition

Water Conservation Programs – A Planning Manual

Project Manager: Melissa Valentine
Production: Studio Text
Cover Art: Melanie Yamamoto
Manuals Specialist: Sue Bach

Library of Congress Cataloging-in-Publication Data

Names: Maddaus, Michelle L., author. | Maddaus, William O., author. | Maddaus, Lisa A., author. | American Water Works Association, issuing body.
Title: Water conservation programs : a planning manual / by Michelle L. Maddaus, William O. Maddaus, Lisa A. Maddaus.
Other titles: AWWA manual ; M52.
Description: Second edition. | Denver, CO : American Water Works Association, [2017] | Series: AWWA manual ; M52 | Includes bibliographical references and index.
Identifiers: LCCN 2017028390 | ISBN 9781625762139
Subjects: LCSH: Water conservation.
Classification: LCC TD388 .M325 2017 | DDC 333.91/16--dc23 LC record available at https://lccn.loc.gov/2017028390

ISBN-13 978-1-62576-213-9
eISBN-13 978-1-61300-413-5

Printed in the United States of America
American Water Works Association
6666 West Quincy Avenue
Denver, CO 80235-3098
awwa.org

Contents

Figures

This page intentionally blank.

Tables

Foreword

This publication is the second edition of the American Water Works Association Manual M52, *Water Conservation Programs – A Planning Manual.* The manual provides information on how to develop, implement, and measure the success of a utility conservation program.

This manual is intended for use by water utilities that are contemplating the development of a conservation program. In addition, water suppliers that already have a conservation program can use this information to make improvements and to gain the benefits of a more comprehensive approach.

With the development of this second edition, not only has the material throughout the manual been updated, but three chapters (5, 6, and 7) and a robust appendix full of case studies have been included. Chapter 5 presents the process of developing a utility water conservation plan, including an example outline for the plan. Chapter 6 aims to foster an understanding of the importance of public participation in developing publicly supported programs as well as building a group of community advocates and partners. Chapter 7 describes the implementation and evaluation of ongoing efforts of a utility conservation program and plan. Appendix A contains 13 case studies representing utilities across the US and Canada, sharing their various experiences with regards to conservation programming, hotel water use, water loss management, greywater reuse initiatives, and other topics.

The Water Conservation Division and its working committees welcome feedback on the content of this manual. As this field evolves, adjustments to the manual will be necessary. Reader comments are a critical part of the review and revision process.

This page intentionally blank.

Acknowledgments

This second edition of the AWWA Manual M52, *Water Conservation Programs – A Planning Manual,* was approved by the Water Conservation Division. Following are the primary authors:

Michelle L. Maddaus, Maddaus Water Management Inc., Danville, Calif.
William O. Maddaus, Maddaus Water Management Inc., Danville, Calif.
Lisa A. Maddaus, Maddaus Water Management Inc., Sacramento, Calif.

The authors acknowledge the significant contribution of the following AWWA Manual M52 subcommittee members who provided case studies, information, and review of individual chapter content:

Veronica Blette, USEPA, Washington, D.C.
Chris Brown, CBConserve, Sacramento, Calif.
Mary Ann Dickinson, Alliance for Water Efficiency, Chicago, Ill.
Eleni Hailu, Los Angeles County Waterworks Districts, Alhambra, Calif.
Bryan Mantz, Public Resources Management Group, Maitland, Fla.
Jeffrey Ripp, Public Service Commission of Wisconsin, Madison, Wisc.
Brian Skeens, CH2M, Atlanta, Ga.
Ann Witmer, University of Illinois, Champaign, Ill.

The following individuals on our committee provided case studies:

David Bracciano, Tampa Bay Water, Tampa Bay, Fla.
Marie Cefalo, Town of Cary, N.C.
Heather Dalrymple, City of Austin, Texas
Laurie D'Audney, Fort Collins Utilities, Fort Collins, Colo.
Julie Friedman, City of Sacramento, Calif.
William Granger, City of Sacramento, Calif.
Drema Gross, City of Austin, Texas
Mialee Jose, Seattle Public Utilities, Seattle, Wash.
George Kunkel, Kunkel Water Efficiency Consulting, Pittsburgh, Penn.
Kelly O'Rourke, Seattle Public Utilities, Seattle, Wash.
Amy Peters, City of Abbotsford, B.C.

Thank you to Jeff Ripp, Dave Bracciano, Chris Brown, Kevin Reidy, and Veronica Blette, who took on the difficult task of reviewing the entire manual and providing detailed comments. The authors carefully considered the reviewers' comments. Selected sections were rewritten by the authors to reflect these comments. This manual represents consensus of the Technical and Educational Council and the Water Conservation Division and its committees.

The following were Water Conservation Division members at the time of approval:

Brian Skeens, CH2M, Atlanta, Ga.
Chris Brown, CBConserve, Sacramento, Calif.
Margaret Hunter, American Water, Voorhees, N.J.
Lisa Krentz, Hazen and Sawyer, Tampa, Fla.
Glen Pleasance, Regional Municipality of Durham, Durham, Calif.
Kevin Reidy, Colorado Water Conservation Board, Denver, Colo.
Lisa A. Maddaus, Maddaus Water Management Inc., Sacramento, Calif.
Veronica Blette, USEPA, Washington, D.C.

The Water Conservation Division wishes to thank the following individuals for their contributions:

Wayne Galliher, City of Guelph, Ont. (case study)
Vlada Keniff, New York City Department of Environmental Protection (case study)
Kathy Nguyen, Cobb County Water System, Ga. (case study)
Donna Scholl, Waukesha Water Utility, Wisc. (case study)
Amy Vickers, Amy Vickers & Associates, Amherst, Mass.

The authors and workgroup members wish to thank Alex Gerling (AWWA staff) for her efforts to bring this manual to print. Molly Beach and Sue Bach (AWWA staff advisors) also assisted during manual preparation.

Abbreviated Terms/Acronyms

af	acre-foot/acre-feet
AMI	advanced metering infrastructure
AMS	advanced metering system
ANOVA	analysis of variance
ANSI	American National Standards Institute
ASME	American Society of Mechanical Engineers
AWWA	American Water Works Association
AwwaRF	American Water Works Association Research Foundation
BMP	best management practice
CalWEP	California Water Efficiency Partnership
CARL	current annual real losses
CII	commercial, industrial, institutional
COM	commercial
CUWCC	California Urban Water Conservation Council
DEP	Department of Environmental Protection
DMM	demand management measure
DSS model	demand-side management least cost planning decision support system
EPA	Environmental Protection Agency
ET	evapotranspiration
FTE	full-time equivalent
GIS	geographic information system
gpcd	gallons per capita per day
gpd	gallons per day
gpd/a	gallons per day per account
gpf	gallon per flush
gpm	gallon per minute
HECW	high-efficiency clothes washer
HET	high-efficiency toilet
HEU	high-efficiency urinal
HOA	homeowners association
ILI	infrastructure leakage index
INST	institutional
IRR	irrigation
IWA	International Water Association
LEED	US Green Building Council Leadership and Excellence in Environmental Design
MG	million gallons

mgd	million gallons per day
MOU	memorandum of understanding
MWM	Maddaus Water Management
NRW	non-revenue water
O&M	operations and maintenance
PG&E	Pacific Gas and Electric
RMF	residential multifamily
RSF	residential single-family
TWDB	Texas Water Development Board
UARL	unavoidable annual real losses
ULF	ultra low flow
ULFT	ultra-low-flow toilet
USEPA	United States Environmental Protection Agency
WEF	Water Education Foundation
WRF	Water Research Foundation

Chapter 1

Introduction

Water conservation is a key component of overall water resources planning. Conservation programs that are carefully designed and implemented can bring many benefits. Among these are the efficient utilization of available sources of supply, public recognition and participation, and improved support for water pricing adjustments.

WHAT IS WATER CONSERVATION?

If this question was asked of water professionals in the 1930s through the 1960s, most would have said "water conservation involves building a reservoir to capture runoff that would otherwise be wasted by flowing into an unusable water body, like the ocean." Starting in the 1970s, water professionals became aware that minimizing water waste was essential. For example, AWWA's water resources policy in 1975 included the statement, "Every effective means to prevent and minimize waste and promote wise use should be employed by all entities, public and private, engaged in water resource activities." AWWA's first handbook on water conservation, *Water Conservation Management,* was published in 1981 followed by an updated handbook, *Water Conservation,* in 1987 (AWWA 1981; AWWA 1987).

The first edition of this manual published in 2006 was an update of these original works (AWWA 2006). This second edition, published in 2017, is a further update with revised case studies, two more chapters, and additional content.

Shown in Figure 1-1 is AWWA's policy on water conservation. The most up-to-date policy statement can be found at www.awwa.org/about-us/policy-statements.aspx. A more extensive description of the role of water conservation in water resources management is contained in AWWA M50, *Water Resources Planning* (AWWA 2017).

Long-Term Versus Short-Term Conservation

A common and long-held public perception is that water conservation means restricting or curtailing customer use as a temporary response to drought. Today, while water use restrictions are still a useful short-term drought management tool, most utility-sponsored water conservation programs emphasize lasting long-term improvements in water use efficiency while maintaining quality-of-life standards. In a growing number of communities, this now includes permanent water use restrictions such as limiting the number of

Adopted by the Board of Directors Jan. 27, 1991, revised Jan. 31, 1993, and June 15, 1997, and reaffirmed Jan. 20, 2002. This policy statement was approved by the Administrative & Policy Council on Sept. 28, 2007, and the Executive Committee on Nov. 3, 2007. Reaffirmed January 2011. Revised June 8, 2014. The most up-to-date policy statement can be found at http://www.awwa.org/about-us/policy-statements.aspx.

The American Water Works Association (AWWA) encourages water and wastewater utilities to support and adopt codes, policies, procedures, and programs that sustain the efficient use of their water resources, through a comprehensive approach integrating demand management with supply-side management practices.

AWWA supports the use and integration of a variety of water conservation principles and practices. Utilities should use comprehensive integrated resource planning to fully utilize conserved water in water supply planning, and participate in regional coordination and integration approaches for this use.

Utilities should accurately meter, monitor and track consumption data and bill customers based on accurately metered usage in a timely manner, using full-cost pricing and conservation rate structures.

Utilities should define and implement cost effective/beneficial best water conservation practices for customer end uses, including the conversion of high-water-use plumbing fixtures, appliances and equipment to more efficient designs; the promotion of efficient irrigation; and the use of appropriate landscape materials. Utilities should also implement cost-effective services which may include the provision of water use audits and technical assistance and the provision of education and public information on local water resource issues and water conservation.

Utilities should also support the development of appropriate new water efficiency certification programs and standards for plumbing fixtures and appliances. Utilities should encourage appropriate on-site water reuse, including rainwater and stormwater harvesting and the use of air conditioner condensate and gray water, consistent with the protection of public health. Utilities should also support continued research into efficient water use practices.

Finally, water and wastewater utilities can provide leadership for water conservation by adopting policies and regulations that mandate efficient water use practices in all public facilities.

Figure 1-1 AWWA official policy on water use efficiency

hours and days per week permitted for lawn and landscape watering because such measures can realize significant reductions in long-term demand. Water conservation, very simply, is doing more with less, not doing without. There are many published definitions of water conservation and water use efficiency. An Internet search will result in dozens of definitions, each slightly different.

Legal Definitions in State Water Codes. Long-term water conservation is often used interchangeably with the terms *demand management* and *efficient use*. For example, the California Water Code states (Section 10611.5):

> Demand management means those water conservation measures, programs, and incentives that prevent the waste of water and promote the reasonable and efficient use and reuse of available supplies.

California goes on to define efficient use separately as (Section 10613):

> "Efficient use" means those management measures that result in the most effective use of water so as to prevent its waste or unreasonable use or unreasonable method of use.

The Texas Water Code (Section 11.002) defines water conservation in two ways:

1. The development of water resources

2. Those practices, techniques, and technologies that will reduce the consumption of water, reduce loss or waste of water, improve the efficiency of the use of water, or increase the recycling and reuse of water so that a water supply is made available for future or alternative uses.

The first part of the Texas definition is based on the earlier concept of retaining water for later use. The second part broadly defines water conservation as including separate concepts of efficiency, reducing use, and recycling water.

The State of Florida defines water conservation [(Section 62- 40.412(1), F.A.C.)] as:

Preventing and reducing wasteful, uneconomical, impractical, or unreasonable use of water resources.

The State of Georgia defines water conservation [SWP (391-3-32-01)] as:

The beneficial reduction of water use, water waste, and water loss.

Georgia offers this guidance relative to preparing water conservation implementation plans:

Water conservation is defined as the beneficial reduction of water use, water waste and water loss. Conservation, implemented as a long-term water management practice, is fiscally responsible and can enhance our ability to grow. Water conservation does not lower our quality of life or deter business. It can lead to more efficient and effective business operations and help water users recognize the value of water. The ultimate goal of water conservation is to maximize the benefit from each gallon used, while not preventing water use.

Dictionary Definitions. The Ecology dictionary (www.ecologydictionary.org/WATER_CONSERVATION) offers the following definition of water conservation:

The physical control, protection, management, and use of water resources in such a way as to maintain crop, grazing, and forest lands, vegetative cover, wildlife, and wildlife habitat for maximum sustained benefits to people, agriculture, industry, commerce, and other segments of the national economy. Water conservation measures result in a reduction in applied water due to more efficient water use such as the implementation of Best Management Practices (BMP): Urban Water Use, or Efficient Water Management Practices (EWMP): Agricultural Water Use. The extent to which these actions actually create a savings in water supply depends on how they affect new water use and depletion.

The Ecology dictionary defines water use efficiency as:

A measure of the crop production per unit of water used, irrespective of water source, expressed in units of weight per unit of water depth per unit area. The concept of utilization applies to both dryland farming and irrigated agriculture.

Water Efficiency and Water Conservation Expert Definitions. Amy Vickers, author of the *Handbook of Water Use and Conservation* (Vickers 2001), presented a paper entitled "Water Conservation or Water Efficiency: What's the Difference?" (Vickers 2008) in which she defined water conservation as a "beneficial reduction in water loss, waste or use," which implies using less or no water. She defined water efficiency, on the other hand, as "minimization of the amount of water used to accomplish a function, task or result." This implies better use of water applied, but not necessarily using less water.

The dictionary definitions suggest that *water use efficiency* is a common term in agricultural use management and *water conservation* applies more to urban water use. Water conservation and water efficiency are often used interchangeably, and the differences are subtle.

For this manual, the following subtle differences will be recognized. Water use efficiency is focused on improvements in water use by means of fixtures and equipment technology. Water conservation focuses on taking action to improve the effectiveness of water use through various means, such as choosing to change behavior by using those fixtures and appliances less. Water efficiency often refers to the state of water use in a community, and higher levels of efficiency implies the community uses less according to metrics such as gallons per person per day or gallons per acre of irrigated area.

WHY CONSERVE WATER?

There are many reasons for water utilities to pursue wise water use and establish a water conservation program. The specific reasons will be different for each utility, and the appropriate level of conservation for a utility should be tailored to local needs. This manual will show utilities how to customize a program to local needs.

There is a broad array of reasons to pursue efficient water use. Some examples for consideration are included below:

- **Cost savings:** Lowering water production and/or distribution costs will save the utility and its customers (or ratepayers) money in reduced operating costs and possibly deferred capital costs. Conservation is often an important part of a least-cost future water supply plan.
- **Wastewater treatment and disposal benefits:** Reduction of indoor water use cuts wastewater flows, resulting in cost savings and lessened environmental impacts of treated wastewater disposal.
- **Environmental benefits:** Water removed from a water body for human use could be used for environmental and other purposes. For example, protection of endangered species often requires a reliable source of good quality water, which might be lessened by water withdrawals.
- **Competing beneficial uses:** In addition to the environment, water left in place could be used for agriculture, power production, recreation, aesthetic enjoyment, wildlife, and so on.
- **Water supply limitations:** Few places now enjoy unlimited water supplies. Water conservation can stretch existing supplies, whether supply is from groundwater or surface water.
- **Avoiding the need for new supply development:** Developing new water supplies is often controversial, and those opposed to supply projects often declare a preference for conservation as an alternative.
- **Utility stewardship and sustainability:** Utilities that conserve water demonstrate leadership in resource management and are working toward a goal of sustainability. The same water resources can sustain enhanced economic activity.
- **Energy savings:** Reducing water production will save energy and reduce greenhouse gas emissions.

- **Improved supply reliability:** Conservation can reduce or postpone drought water use curtailments by essentially increasing supply (i.e., building a drought reserve).
- **Customer benefits:** Customers who conserve water may enjoy lower water bills and possibly lower wastewater and energy bills.
- **Regulatory compliance:** Some state regulatory agencies require water conservation plans and/or implementation progress to qualify for permits, grants, and loans. Some states have set per capita use reduction targets, which implies the need to increase conservation efforts.
- **Public perception:** The public often insists on demonstrating efficient use of existing water supplies before supporting expansion of supplies to meet new water needs.

Considerations to Pursuing Efficient Water Use

Sometimes there are factors that must be carefully weighed before pursuit of aggressive conservation is implemented. Considerations may include the following:

- Reduction of water use requires utilities to modify their demand and revenue forecasts, rates, and/or rate structures.
- Some utilities still are concerned about *demand hardening* – the impact of long-term demand reductions on the ability to reduce demand further in times of droughts or water shortages. Analysis should also consider how reduced demand stretches existing supplies, delaying the need for drought or shortage measures (*curtailment*).
- Some utilities need assistance from specialists in water conservation to provide specific expertise on how to implement conservation programs and properly assess the benefits from such programs. Water utilities are selling the service that water provides to meet customer needs. This is a new focus and perspective for water utilities – to meet customer needs with current revenue levels, rather than just looking at the total number of gallons of water that are sold.
- Many utility billing systems do not support customer sector water use data needs and analysis. Implementing a conservation program may require reprogramming of billing software to support the utility programs.
- In some locations, conservation can threaten the "use it or lose it" doctrine of water law and water rights. Thus, during times when they are water efficient, some senior water rights holders have begun considering the option of selling their excess water to others so as not to "lose" their water rights because of reduced usage.

This manual attempts to give utilities ways to address many of these considerations and to identify potential drawbacks so they can pursue new ways of meeting customers' needs or improving existing methods.

PLANNING AS PART OF A WATER CONSERVATION PROGRAM

In 2013, AWWA released the G480 standard Water Conservation Program Operation and Management (AWWA 2013). The voluntary G-series standards are intended to help utilities improve their overall operation and services. The guidelines include appropriate practices, procedures, and actions that can help promote effective and efficient utility operations.

The key elements of the standard include:

- **Top-level organizational functions:** Staff, planning, education, ordinances
- **Internal utility actions and requirements:** Metering, rate structure, billing practices, landscape efficiency program, distribution system, and pressure management program
- **External policy requirements:** Codes and standards
- **Wholesale agency requirements:** Regulatory requirements, education, internal utility actions, and requirements

The development of a water conservation plan is a critical element of an effective water conservation program and is a requirement to meet the G480 standard, which is also discussed in chapter 5. The actions and requirements that are included to meet the standard will also be evaluated and considered as part of the planning process. Users of this manual are encouraged to review the G480 standard to ensure that they are developing an effective holistic program to manage water demand in their service area.

Steps to Develop a Water Conservation Plan

To start a water conservation program, a water conservation plan should be developed. In 1998, a group of water efficiency experts participated in the development of USEPA's Water Conservation Planning Guidelines (USEPA 1998). The guidelines included ten basic steps to outline the activities undertaken in a water conservation planning effort. Although the structure of this manual does not strictly align with these ten steps, reviewing them clarifies the scope of efforts that need to be considered in developing a cost-effective program. The ten steps focus on the technical content of a water conservation plan, relying on cost-effectiveness analysis to determine the scope of the conservation program (i.e., the measures selected to be implemented). These steps are covered in chapters 3 and 4 of M52. Preparing the written conservation plan involves getting it adopted, funded, staffed, implemented, and monitored (tracking progress versus conservation goals). These topics are covered in chapters 5, 6, and 7 of M52.

1. Review detailed demand forecast
2. Review existing water system profile and descriptions of planned facilities
3. Evaluate the effectiveness of existing conservation measures
4. Define conservation potential
5. Identify new conservation measures or enhancements to existing conservation measures
6. Determine feasible measures
7. Perform benefit–cost evaluations
8. Select and group conservation measures into alternative scenarios
9. Combine overall estimated savings
10. Optimize demand forecasts

Review detailed demand forecast. A baseline forecast of the water use analyses and the demographic (customer account or population) projections should be created. The impacts of current and selected additional conservation measures can be superimposed on the baseline forecast.

Review existing water system profile and descriptions of planned water supply facilities. As water demands increase, utilities need to maintain information necessary to develop and update a system profile from an inventory of existing resources and conditions. A review of this information is essential for accurately targeting water conservation measures (e.g., reductions in peak-day water use) as appropriate emerging needs.

Evaluate the effectiveness of existing conservation measures. If existing conservation measures are present in the water use analyses, the degree of current and prospective conservation stemming from these measures needs to be quantified. Some of this effect could be naturally occurring if it results from code requirements, for example, in the US Energy Policy Act of 1992, which requires that replacement fixtures and fittings in new construction meet water efficiency standards. Forecasts of the overall water savings from naturally occurring conservation measures is about 5 to 15 percent of total water needs by 2030 (CUWCC 2001).

Define conservation potential. A detailed assessment of indoor and outdoor water use by existing and new customers is essential to determine the conservation potential. A comparison of the water use profile with those in industry studies such as *Residential End Uses of Water, Version 2* (WRF 2016) and *Commercial and Institutional End Uses of Water* (Awwarf 2001) should be made to identify the potential for additional conservation.

Identify water conservation measures. Even though many water conservation measures are transferable among locations, water conservation measures should be tailored on a case-by-case basis to develop the most effective program for local conditions within a given service area.

Numerous water agencies around the world, particularly in arid climates (for example the arid parts of the southwestern United States), have been implementing water conservation programs for well over 20 years. General conservation methods, both as internal utility actions and through customer participation, that can be targeted include both basic and more advanced measures.

Basic measures:

- Public education
- Codes and standards
- Water waste restrictions
- Consumption-based metering and billing
- Water distribution system improvements (leakage reduction)
- Metering water use

More advanced measures:

- Irrigation efficiency improvements
- New-home xeriscaping (low water use landscaping)
- Large landscape irrigation improvements
- Residential home water efficiencies
- Large commercial efficiency projects
- Small commercial efficiency projects
- Municipal publicly owned building indoor and outdoor retrofits
- Replacement of inefficient toilets
- Commercial landscape ordinances
- Industrial and institutional efficiency projects
- A rate structure that promotes conservation, such as seasonal rates or water budgets

Most utilities that have not implemented a conservation program will want to look first at the basic measures. After they have some experience, they can proceed to the more advanced measures.

Determine feasible measures. Not all conservation measures will be practically, politically, or economically feasible for a given utility. For example, drought-tolerant landscaping is not suitable for some climates or some utilities; an inclining block rate structure is not suitable for an unmetered area or where there is strong customer resistance; and capital-intensive reclamation facilities will not provide an economic return in smaller communities. To complete the feasibility analysis, the number of accounts that could and would use each measure and the specific savings over time that would accrue to its implementation must be determined. In addition, the existence of legislative or institutional obstacles to implementation needs to be researched. Estimates of market penetration are based on measure design and experience from similar measures implemented by other water utilities.

More than 100 individual conservation measures could be implemented among the residential, commercial, industrial, irrigation, agricultural, and public authority accounts in large metropolitan areas. The implementation of conservation programs usually includes customer education, sometimes financial assistance (toilet rebates), sometimes financial incentives (conservation rates), and sometimes legislation (plumbing codes for ultra-low-flow [ULF] toilet replacements). Measures can be qualitatively screened to a shortlist of the most promising measures. The shortlisted measures can be evaluated for water savings, benefits, costs, and practicality.

Perform benefit–cost evaluations. If supply is severely limited, the benefits of conservation are virtually priceless: It is a matter of having enough water for essential indoor residential and commercial needs. Under less extreme circumstances, however, it is necessary to conduct a basic benefit–cost analysis that relates the value of water saved to the cost of implementing the program over a useful program life. A frequent basis for valuing conservation programs is through the benefits associated with the delay, downsizing, or averting of new facilities. Some communities engage in modest conservation efforts as part of public-spirited programs that link with ecological and environmental goals for a better world to live in. Benefits are often measured from the consumer's point of view, usually in terms of less water consumption to pay for and less energy cost for heating water.

Select and group conservation measures. Individual conservation measures should be formulated into one or more comprehensive programs for further review to decide on the best program for implementation. Each program scenario will include that array of justifiable outdoor and indoor measures that meet the payback criteria and will achieve needed and targeted results. The recommended program scenario must also be acceptable to the utility management and governing bodies to be included in long-range demand forecasts. A stakeholder–public process should be used to confirm or guide the selection of the best alternative program scenario of measures.

Combine overall estimated savings. Once an optimal mix of conservation measures has been determined, an overall estimate of program water savings can be developed with a cautious summation that avoids counting estimated water savings from individual measures more than once (e.g., residential toilet leak water savings and toilet replacement). Also, an overall program implementation schedule for the recommended program of measures is necessary to determine the timing of conservation effects on the demand forecasts.

Optimize demand forecasts. The baseline demand forecast should be modified for quantification of demand reductions and graphical comparison of the water forecast with and without conservation. Modification of demand forecasts may be done iteratively or simultaneously with different cost-effective groups of conservation measures that have been formulated into alternative scenarios to meet desired conservation targets.

Alternative program scenarios could include a mandated savings goal, most cost-effective measures, most water savings, and so on. By integrating anticipated conservation, utilities can avoid rate revenue surprises. By integrating demand with the rate forecasters, utilities should recognize that implementing water conservation is successful.

DEVELOP A WORK PLAN

Responsibilities of the Water Conservation Program Manager

To ensure a successful water conservation program, it is important to have a water conservation coordinator who is responsible for planning and implementing the plan. The responsibilities of the water conservation program manager are, initially, to develop the long-range efficiency plan and then organize and direct the various measures that the recommended program comprises. This begins with preparing a work plan that defines the schedule and budget for each task identified to implement the plan. In a small utility, the manager will work part time on water conservation and be responsible for carrying out most tasks. In larger utilities, managers will have the option of assigning other staff to individual tasks while they coordinate the overall program.

Work Plan

Implementation can be a long, slow process, similar to planning, designing, and building capital facilities. A 10-year time period from implementation to actual water savings benefits may often be appropriate. Many conservation measures take about three to four years to become fully operational.

The following guidelines may help utilities with implementation:

- Establish clear lines of communication for staff and management
- Obtain the necessary funding for selected measures
- Decide on appropriate mix of staff and contractors for each measure
- Consider teaming with neighboring utilities to capitalize on economies of scale
- Hire or assign staff to coordinate each measure
- Design the individual measure setup; actively market the programs to potential participants, which may include advertising or individual contacts (for commercial enterprise, for example)
- Involve elected officials in the launching of each measure
- Involve the public in marketing measures
- Publicize the success of each measure
- Evaluate the cost-effectiveness of each measure
- Update the efficiency plan every two to three years

Examples of implementation tasks for specific measures may include:

- Developing a public information and in-school education program
- Setting up and conducting speakers' groups with volunteer or paid presentations about the water efficiency program
- Disseminating information and conducting public education activities
- Supervising retrofit device or fixture distribution

- Offering assistance to large users of the system such as industries, universities, parks, and so forth
- Overseeing the utility water loss control and leak reduction program
- Revising local laws, codes, or ordinances to require the installation of water-saving fixtures
- Developing incentives to encourage efficiency, including appropriate water pricing and rebates
- Liaising and coordinating with programs run by neighboring water supply utilities

Responsibility of Program Participants

In addition to the water conservation program manager, other individuals and groups may be involved in program implementation. These persons or groups and their roles include:

- The water utility manager approves the final efficiency plan and authorizes budget and staffing requests. The manager will also extend formal requests for participation on a water efficiency advisory committee, if desired.
- The water utility board of directors, whose members may be publicly elected, is often supportive of water conservation programs, as such programs are popular with customers and public interest groups. The efficiency program manager should use all possible opportunities for presenting success stories at board meetings to advocate the authorization of additional programs and funding.
- The water efficiency advisory committee. Medium-sized and large utilities often have an advisory committee, the role of which is to review and comment on plans, potential measures, and implementation strategies. The committee can either be internal or a citizen advisory committee.
- Consultants specializing in developing efficiency plans, providing advice on the implementation of measures, and evaluating water savings and cost-effectiveness resulting from completed measures.
- Contractors, who are sometimes hired to conduct programs.
- Public information specialist. Special skills are required to handle the program aspects related to publicity and public education. In the beginning and periodically, this specialist can direct the stakeholder–public process used to help select the plan and periodically update it. The task can be implemented in house or contracted to a public relations company.
- Participants. The program will not succeed without the participation of targeted customers. They should be encouraged, with an offer that is too attractive to decline, to participate in making the changes in order to achieve efficiency. Education, regulations, and incentives such as rebates can all convince customers to participate.

PARTNERSHIP OPPORTUNITIES

One way to prepare and implement a conservation plan is to work with another water utility or city/county department during the planning process. This can reduce plan preparation and implementation costs and make for more effective programs. Opportunities to partner include the following:

- **Integrated resource plans** (IRPs) can be prepared to address future supply and demand imbalances. This is probably the best way to address the role of conservation in utility water resources planning. IRPs often identify stakeholders and potential partners.
- **Wholesale water agencies** may offer an opportunity to partner on planning and implementing water conservation programs. Often the wholesale agency can implement certain elements of the plan more cost-effectively. Preparation of regional educational materials, regional promotions, discount purchases, and purchase of media time are examples of getting the wholesale agency involved. Often funding this approach is a matter of amending one master agreement.
- **Energy, stormwater, solid waste, and wastewater utilities** sometimes have common interests relating to water conservation. Oftentimes, partnerships increase the likelihood of program implementation. Examples of water and energy partnerships abound, such as promoting water-efficient clothes washing machines that save both water and energy. Wastewater agencies may be interested in jointly funded flow reduction programs, such as a toilet rebate program, if it can help them meet discharge requirements or capacity constraints. These partnerships are more likely to help in the area of program implementation, rather than in plan preparation.
- **City/county departments**, such as planning, building, and parks departments, can work with a water utility to achieve common goals, including saving water, energy, and money.

ORGANIZATION OF MANUAL

The manual is organized into seven chapters and four appendixes.

- **Chapter 1 – Introduction:** The first chapter sets the stage for what follows by defining water conservation, listing reasons to conserve water, and stating AWWA's policies and positions on efficient water use.
- **Chapter 2 – Understanding Conservation and Setting Goals:** This chapter describes how to assess the need for conservation and describes in more detail the pros and cons of establishing a program. A process to set conservation program goals, qualitatively and quantitatively, is proposed. Coordination of utility goals with other department and agency goals is described. The goals of leading organizations focused on water conservation are highlighted.
- **Chapter 3 – Analysis of Water Use and Water Savings:** Chapter 3 provides a method to forecast future water needs before additional conservation is considered. Methods to assess and reduce water losses are covered, and a list of customer conservation measures is presented. The reader is referred to various sources for cost and savings data on conservation measures. A screening process can be used to reduce the number of measures to be quantitatively evaluated.
- **Chapter 4 – Evaluation of Benefits and Costs:** This chapter describes how to estimate water savings from conservation measures, benefits, and costs and how to find the ratio of benefits to costs so as to be able to compare cost-effectiveness of alternate conservation programs. Different perspectives on benefit–cost analysis are offered, and an example calculation is provided.
- **Chapter 5 – Creating a Formal Water Conservation Program Plan:** Once future water needs have been assessed, options to conserve water have been analyzed,

and goals set, the recommended plan can be selected and details including schedule and budget can be established.

- **Chapter 6 – Stakeholder Involvement:** It is important to have the support of key stakeholders when developing, implementing, and adopting a plan. This chapter describes how to involve the public in developing and carrying out the program.
- **Chapter 7 – Plan Implementation, Managing Revenue Impacts, and Evaluating Performance:** Failure to adapt to changing circumstances can hinder the success of a conservation program. This chapter provides information on implementation issues, including balancing the need for conservation with utility revenue needs. It also provides details on monitoring and evaluating plans with discussion of review cycles as triggers for when plans should be revisited. Metrics and reference lists are provided for use in evaluating the efficiency of different types of facilities (e.g., gallons per hotel room, gallons per square foot of office space).
- **Appendixes:** Appendix A includes several case studies from water utilities highlighting successful implementation of conservation measures. Appendix B provides a checklist for the AWWA G480 standard for Water Conservation Program Operation and Management. Appendix C provides guidelines unique to conservation for involving both internal stakeholders across the utility (finance, operations, public affairs, and planning/engineering department staff) and externally across the community (business, residents, etc.). Appendix D provides details related to data management (e.g., customer service, operations and finance) necessary to monitoring and evaluating conservation program long-term effectiveness (may also aid during short-term drought emergencies).

REFERENCES

American Water Works Association Research Foundation (Awwarf). 2001. *Commercial and Institutional End Uses of Water.* Denver, CO: Awwarf.

American Water Works Association (AWWA). 1981. *Water Conservation Management.* Denver, CO: AWWA.

AWWA. 1987. *Water Conservation*. Denver, CO: AWWA.

AWWA. 2006. *Water Conservation Programs – A Planning Manual,* 1st ed. Manual M52. Denver, CO: AWWA.

AWWA. 2013. Water Conservation Program Operation and Management. Standard G480. Denver, CO: AWWA.

AWWA. 2017. *Water Resources Planning,* 3rd ed. Manual M50. Denver, CO: AWWA.

California Urban Water Conservation Council (CUWCC). 2001. "Impact of the National Plumbing Efficiency Standards on Water Infrastructure Investment." *Annual Report 2001.* Sacramento: CUWCC.

US Environmental Protection Agency (USEPA). 1998. *Water Conservation Plan Guidelines.* EPA-832-D-98-001. Washington, DC: USEPA.

Vickers, Amy. 2001. *Handbook of Water Use and Conservation.* Amherst, MA: WaterPlow Press.

Vickers, Amy. 2008. "Water Conservation or Water Efficiency: What's the Difference?" WaterSmart Innovations Conference, Las Vegas, Nevada, October 9, 2008.

Water Research Foundation (WRF). 2016. *Residential End Uses of Water, Version 2*. Denver, CO: WRF.

This page intentionally blank.

Chapter 2

Understanding Conservation and Setting Goals

This chapter presents an overview of typical water conservation programs and incentives that water utilities undertake, and how these programs can be designed to achieve goals for multiple purposes as defined by the water utility. The building blocks for designing conservation programs are provided in this chapter: It discusses the overall need for conservation programs, outlines the iterative process of setting qualitative goals, describes the typical incentives that exist for promoting conservation, explores potential advantages and disadvantages, defines the potential benefits of partnerships with other local utilities or public works departments, and gives an overall approach to planning and implementing conservation programs.

Defining the appropriate level of water conservation for a water utility is a complex process. It requires that basic goals be identified that will direct the planning and analysis of conservation options, in order to help choose appropriate strategies that are cost-effective, as discussed in chapter 4, and that will shape the overall long-term efficacy of the conservation programs for the utility. Goals focus the conservation program and shape tracking implementation efforts. The key to success is setting the right goals at the outset in a clear and transparent process. Transforming initial goals into cost-effective water conservation targets is essential, and further refining the targets and goals throughout the implementation phase will result in higher-value conservation benefits in the long run.

TYPICAL CONSERVATION DRIVERS FOR WATER UTILITIES

Water conservation is a necessary component of a successful overall water supply plan. Utilities undertake water conservation programs to (1) reduce water and wastewater demand and capacity utilization and operating expenses, thus achieving significant cost savings; (2) meet regulatory requirements; (3) meet long-term water supply planning

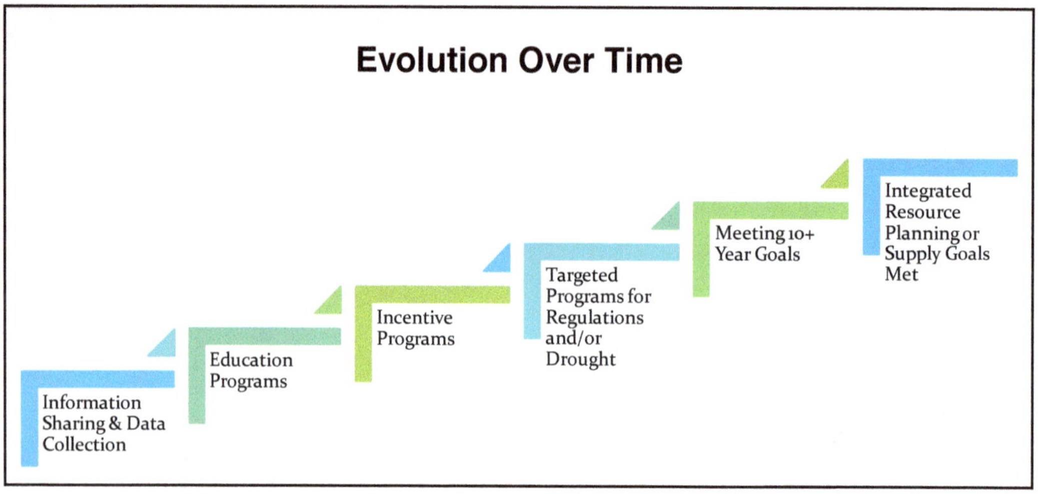

Figure 2-1 The evolution of water efficiency drivers

needs; (4) improve customer service and lower customer costs; (5) reduce greenhouse gases; and (6) address short-term emergencies (or curtailment needs). Efficiency in a system is achieved when demand is decreased to optimize available or planned future water supplies and treatment capacity. When investigated as part of the integrated resources planning (IRP) process, conservation may serve as a complementary partner to water supply planning or as a stand-alone supply (deferral) alternative.

Figure 2-1 shows an evolution of drivers for water conservation over time. As the program matures, more innovative programs are implemented and goals are set, and water conservation becomes a source of new water supply. These concepts are further described in AWWA Manual M50, *Water Resources Planning*. Note that drought emergencies and other short-term shortages are covered in AWWA M60, *Drought Preparedness and Response*. Long-term shortages and the need to plan for them are the overall focus of this manual.

Figure 2-2 shows a new home built without a front lawn, thereby increasing overall household water use efficiency.

In the United States, federal and state agencies encourage conservation through implementation of laws and regulations, codes, and standards; development of guidelines; sponsorship of research; dissemination of information to water utilities; and assistance with program funds through grants or loans. For example, the US Department of Energy oversees the National Plumbing Efficiency Standards, which mandate water efficient plumbing fixtures. In addition, many states require a water conservation plan before a new water right or permit of use is granted or as a condition of financial assistance. Some states mandate per capita water reductions on water utilities and their customers.

Technology has changed in the United States over time. Figure 2-3 shows a very old toilet that is no longer in use.

Long-Term Planning Considerations and Issues

AWWA Manual M50, *Water Resources Planning*, along with many other leading resources on IRP, addresses the complex matrix of long-term planning issues that face a water utility. Certain themes focused on water conservation are repeated in many of these references, for example:

- Rates of increased demand can be slowed by conservation, particularly on a per capita basis, even though demand can increase overall because of population growth.

Figure 2-2 New home built without front lawn
Source: Maddaus Water Management (MWM)

Figure 2-3 Old toilet technology
Source: Maddaus Water Management

- Conservation will reduce both the average and the peak demand.
- Conservation will decrease operation and management (O&M) expense (see chapter 4) as well as capital expenditures.
- The full benefits of conservation can be measured only by including the financial value of deferring capital expenditures.
- In addition to downsizing or avoiding water supply costs, some wastewater capital expenditures can also be deferred. These benefits should also be counted as avoided costs.
- Geographic differences need to be considered in estimating avoided conservation cost.
- Environmental externalities and energy savings must be evaluated.
- Long-term planning goals should be developed in a stakeholder process with utility managers, elected officials, and customers.

These themes should be incorporated into the qualitative goals of a water efficiency program.

GOALS FOR CONSERVATION

The level of conservation needed will be unique to each utility, and assessing that need should be done in a stakeholder involvement process that includes utility personnel, public officials, customers, and regulators and is designed to brainstorm the key reasons to undertake a conservation program. To assist in the brainstorming process, the following section outlines some key questions and some potential advantages and disadvantages for pursuing water conservation alternatives.

Setting Goals

Developing appropriate water demand reduction goals is an essential step to properly designing water conservation programs. Goals provide a yardstick against which to measure progress in reducing consumption, and should be set by local water utilities and their stakeholders at the outset. Goal setting is often primarily based on the need to save water, taking into consideration the regionwide ability to extend available water supplies and prevent capacity shortages in individual utility water systems, as well as to provide important environmental benefits to the watershed as a whole. Goals can be set on the supply side as well as the demand side, and for other general purposes.

On the supply side, answers to the following questions will help to formulate specific goals for individual utilities:

- If the system has a water supply shortage, is it limited to one utility or one portion of the service area, or is it a regionwide shortage?
- Is the supply constrained by a drought emergency or long-term (multiyear) occurrence?
- Is the shortage occurring now or is it projected to occur in the future?
- What is the primary cause of the long-term supply shortage? Possibilities could include high growth in demand, system leaks, withdrawal permit limits, transmission mains or water treatment plant capacity issues, distribution pipeline delivery limitations, or inadequate water supply.
- Is the shortage related to water or wastewater treatment capacity?

- Does the supply or treatment shortage occur during peak demand periods each day, during the high-use seasons of the year only, or does it occur throughout the year?

On the demand side, answers to the following questions will help to set a quantifiable goal to reduce demand:

- What level of water use reduction is needed over what time frame? (Typically a 5–10 percent reduction could be considered small and can usually be achieved within 3–10 years, 10–20 percent is a medium goal that may require 10 or more years, and 20 percent or more is a large reduction or an ambitious goal that should be set over a longer period such as 15 or more years.)
- Is there a per capita water use target that has been set by an outside entity like a regulator that must be met?
- What class of customers will be impacted the most by conservation?
- What categories of water use are growing the fastest?
- Where is the most obvious conservation potential?
- How much can be saved in a cost-effective manner?
- How will the revenue impacts from the reduced sales be defined and integrated?

General overall goals can be expressed in terms of:

- Total water savings at some point in the future, expressed as a percent of total production and/or quantity of water saved
- Benefits realized, such as a major project deferred or avoided and water made available for future growth
- Lowered per capita consumption, to reflect the community's balance with the environment, or to meet regulatory or environmental constraints

Specific goals for conservation measures should be developed. Goals that measure implementation progress in terms of specific activities, such as the number of audits completed in a year, can be useful for monitoring progress. This information is easier to acquire and track and is generally available before water savings can be accurately measured.

Measuring progress against goals is useful to the utility to ensure continued program support and funding; to state and federal regulatory or permitting authorities to show that water is being used efficiently and progress can be ascertained; and to adaptively manage goals, as necessary. Measuring savings is an important part of measuring progress, and chapter 3 of this manual describes methods of evaluating savings on an individual measure basis. Water savings for some measures, such as public information, are not easily quantified because savings may be intertwined with other measure effects, and data to separate the effects is rarely available. Water savings from other measures, such as plumbing fixture retrofits, are available because reliable water savings data has already been published.

Defining Program Benefits and Constraints

By defining the clear benefits to the utility and identifying any potential constraints, the planner can establish a tailored set of *qualitative* goals. This is an important first step. Once established, the utility planner will be better organized when assessing the *quantitative* goals for water savings, as presented in chapter 3. In addition, understanding the benefits

Table 2-1 Overview of Benefits From Water Conservation

Recipient of Benefit	Type of Benefit	Description
Water utility	Supply system operations and maintenance (O&M)	Short-term and long-term O&M costs reduced as a result of lower energy expenses because of reduced pumping and chemical use in water treatment and disposal. (The Electric Power Research Institute estimates that from 4% to 5% of all electricity used in the US is used for pumping water.) For city governments, energy usage for water and wastewater utilities can exceed 60% of total energy expenses.
Water utility	Supply system capital investment	Capital facilities for storage and treatment can be deferred or downsized.
Water utility	System reliability	Less water purchased from wholesale water purveyors and more reliability of supply yield, depending on capacity availability.
Wastewater utility	System O&M	Short-term and long-term O&M costs reduced as a result of lower energy expenses because of reduced hydraulic loading on collection systems, volume pumping, aeration, and chemical use in wastewater treatment.
Wastewater utility	Disposal system capital investment	Capital facilities for land disposal can be deferred or downsized. Additional benefits when wastewater discharge restrictions exist.
Environment	Quality enhancement	Decreased need for dams and reduced construction disturbance in natural waterways.
Environment	Quality enhancement	Higher instream flows for fish and wildlife habitat and lower withdrawals from supply sources.
Environment	Quality enhancement	Lower discharges of treated wastewater to receiving waters.
Environment	Quality enhancement	Reduced pollution because of less construction of capital facilities.
Community	Aesthetic quality	Diminished aesthetic effects on waterways from avoided capital projects.
Community	Environmental justice	Fewer social equity issues with facility concerns.
Community	Economic	Increased economy on the same resource, creation of water conservation jobs, and customer savings in utility bills for both water and energy.
Community	Economic/political	Fiscal savings from avoided or delayed new capital expenditures or bond indebtedness.

Source: Adapted from AWWA M50, *Water Resources Planning*, 2nd ed.

of the conservation program can help refine approaches to implementing measures that are targeted at end uses of water that have the highest probability of gaining desired water savings.

Potential Benefits. The benefits of water conservation apply to utilities of all sizes located throughout the United States and abroad. An overview of general benefits of water conservation programs is summarized in Table 2-1.

Potential Constraints and Perceptions. Relying on conservation measures may also pose some potential constraints for utilities or solicit negative perceptions from stakeholders. Some of these issues are summarized along with potential solutions in Table 2-2.

The process of reviewing both the benefits and constraints will be valuable as the goals become defined and further refined.

Mutual Goals With Other Utilities or Departments

There is great benefit and synergy to working with other local utilities or internal public works departments in putting together a conservation program. A brief summary of each type of organization and their potential interest in water conservation programs follows.

Wastewater utilities: There are two primary benefits to wastewater systems from the reductions of wastewater volumes caused by decreases in indoor water use from more efficient plumbing fixtures and appliances. Most notably, benefits are in the form of (1) reductions in O&M costs (energy, chemicals, land disposal); (2) assistance in meeting waste discharge regulatory requirements that are volume based and, in some limited

Table 2-2 Overview of Potential Constraints and Perceptions

Issue	Perception	Potential Solution
Lack of customer participation/ understanding	Customer's view: conservation vs. rationing (sacrifice only needed in times of drought).	Educate customers on their water use and opportunities for everyday cost savings on energy and water bills, such as looking at life-cycle costs of more efficient clothes washers.
Conserving water reduces revenues	Lower water sales results in less revenue for the utility.	Educate customers, utility management, and political leaders on how conservation can reduce system costs. Examine rate structures for possible needed adjustments. Evaluate adjustments needed to raise capital for the investment in the next water supply project. The cost savings from avoided operations and maintenance (O&M) costs and deferred capital facilities are directly accounted for as a benefit to the demand management program (see chapter 4).
Expense of water conservation program	Efficiency programs, particularly rebate incentive programs, are not always cost-effective.	Evaluate demand management measures, as described in chapter 4, for level of cost-effectiveness prior to implementation.
Ineffectiveness in achieving desired water savings	Estimated water savings do not develop.	If water savings do not develop, reevaluate program design (e.g., delivery mechanisms, community attitudes, and recent successes of other utilities) and adjust as needed. Water savings estimates continue to be refined with new information and technologies becoming available. Quantifying water savings with best available information is necessary, as discussed in chapter 5.
Short-term drought savings become more difficult	Conservation measures over the long term reduce the available amount of water savings. Otherwise known as demand hardening, may limit a future drought response in the ability of customers to save water.	Evaluate the role of long-term conservation programs in reducing the negative impacts of both the frequency and severity of drought.
Capital investment up front required to start programs	Water savings accrue over time with avoided costs to the utility, but the investment is required at beginning of the program.	Avoid disjointed investments ($5 million one year and none the next) and spread the investment out over multiple years. Seek grant or loan funding to assist with subsidizing program costs and evening out the cost outlay to match the cost savings.
Impact on water quality	Water aging due to longer residence times in the distribution system may mean more system flushing to preserve chlorine residual.	Evaluate residence times throughout the system and devise appropriate measure to control pathogen growth. Conservation is not the only cause of water aging. Other causes are (1) slower growth planned for an area so infrastructure is oversized, (2) lower peak day flow due to less landscaping water use, (3) use of recycled water for irrigation rather than potable water, (4) urban development that increases longer transit times between water treatment plant and end uses, and (5) effects of drought that permanently reduce water use in an area where infrastructure has already been installed.
Impact of lower flows on drain lines	Reductions in indoor water use can create insufficient drain line flow, particularly in commercial buildings.	Evaluate if any drain line blockages have been reported in green buildings. Consult the Plumbing Efficiency Research Coalition's Report on this issue (PERC 2016).
Impact of lower flows on wastewater treatment plants	Wastewater treatment operations may need to be managed differently for load.	Evaluate whether reduced flows would cause problematic loading at the wastewater treatment plant. Most load issues can be resolved through careful management. Excessive hydraulic capacities often need to be addressed through a new and expanded facility, a problem that can often be addressed or delayed through conservation.

cases, concentration based (e.g., reductions in oil and grease from limiting household garbage disposal use).

Energy utilities: The benefits from less use of heated water (e.g., water efficient showerheads and clothes washers) directly translate into energy savings for energy utilities. Many energy utilities already co-fund programs with local water utilities to capture these mutual benefits. An example of this co-funding is the Pacific Gas and Electric Company's joint program for high efficiency clothes washer rebates undertaken with its local water utilities (PG&E). But some electricity providers fund programs beyond just saving the heating of hot water. Southern California Edison has funded leak detection and water loss management programs in some of its high energy use water utilities to avoid leaking water that has a high embedded energy cost (SCE).

Stormwater: Reductions in runoff translate into decreases in pollutant loading of nutrients from fertilizers and pesticides through stormwater systems in summer months. Conservation programs that make outdoor landscape watering more efficient will reduce pollutant-laden runoff and provide an environmental benefit to receiving streams. Opportunities for rain catchment systems in some areas may be an option for irrigation with nonpotable supplies.

Solid waste: Overwatering of landscapes produces more green waste that needs to be taken to landfills. The primary benefits of solid waste reduction are seen in decreases in O&M (trucking) costs and landfill space. However, if bioreactors, which convert green waste into energy, are used, reductions in green waste will not be a driving factor for water conservation programs.

Health departments: Principal benefits are found in leakage reduction programs, which lead to repairing leaks that further lowers the risk of contamination into the distribution system. Demand reductions also allow utilities that use multiple sources to optimize and use the sources with the highest water quality first.

Environmental organizations: Other organizations include those focused on environmental quality, such as aquatic ecosystems or wildlife habitat, that seek improvements in water quality or time and place of water withdrawals or the quantity and water quality of wastewater discharges.

Professional nonprofit organizations: Organizations such as the American Water Works Association, Water Environment Federation, and the American Society of Civil Engineers track the status and research needs of the water industry, including efforts to promote water conservation.

Other nonprofit organizations: Numerous nonprofit organizations assist with educating the public on health, environmental, water security, and other concerns. These organizations typically conduct research that may have interest in partnering in demonstration projects with water utilities. Such organizations that have national recognition include the Alliance for Water Efficiency, US Green Building Council, Rocky Mountain Institute, Global Energy Partners, Consortium for Energy Efficiency, and American Council for an Energy-Efficient Economy. Other state or local nonprofit organizations exist, such as the California Water Efficiency Partnership and local watershed groups that may have interest in partnering on water conservation programs. The Water Research Foundation (WRF) has also funded research and publications on water conservation.

Park departments: National, state, and local parks can team with water utilities and/or take on water conservation measures on their own, like installing water efficient fixtures and landscape.

Federal agencies: The US Environmental Protection Agency (USEPA), Office of Water, assists with disseminating information, funding grants, and developing guidelines for water conservation planners, as well as administering the WaterSense product labeling program (www.epa.gov/watersense/). Partnered with the USEPA, the US Department of Energy oversees both regulatory requirements for energy and the Energy Star labeling program (www.energystar.gov), which promotes water efficient plumbing fixtures and appliances. The US Bureau of Reclamation has an active role, particularly in the western United States, with promotion of water conservation requirements in their contracts, reporting requirements on implementation, and funding assistance. The Federal Energy Management Program (FEMP), US Army Corps of Engineers, and the Department of Housing and Urban Development have interest in demand reductions at military facilities and federally owned housing projects.

ALTERNATIVE APPROACHES TO TARGET SETTING

Should the utility find it advisable to set targets or specific goals on which to gauge whether utilities are performing up to expected levels, the following alternatives could be considered.

- Sector target approach: This could take the form of
  - a per capita water use target for residential use
  - a per-account target for nonresidential use
  - a volumetric or Infrastructure Leakage Index (ILI) target for leakage reduction
- Best practices program approach: The two options include different methods of agreeing activities and are described as follows:
  - Option A: Develop a memorandum of understanding (MOU) comprising a range of activities (best management practices [BMPs] or reasonable conservation measures [RCMs]), through a stakeholder consensus process, similar to the California approach. Penalties for noncompliance would be sufficiently harsh to ensure performance.
  - Option B: Mandate, using a regulatory framework, a requirement to develop and implement specific activities (BMPs or RCMs). This would be similar to the current Arizona Non-Per Capita Conservation Program discussed later in this chapter. See sidebar "State of Arizona Municipal Water Use Requirements."
- **Hybrid approach to target setting:** If targets must be set, a hybrid approach based on the UK leakage targets (Strategic Management Consultants 2012) and the California BMP and MOU consensus model can be considered. The utility would benefit from direct involvement of stakeholders in the process of developing goals, an MOU, and/or BMPs. A hybrid approach could involve the following:
  - Water efficiency program: Include milestones involving specified actions and schedules that could be termed water efficiency activities (WEAs) for the residential and business sectors. The reporting should follow the key details of the WEAs (or BMPs) and should include the following:
    - Planned versus actual activities
    - Planned versus actual expenditures
    - Estimated water saved
    - Any variations made (if allowed) to implement the program in a manner that is at least as effective as originally planned. Reasons for variations should be justified.
  - Leakage reduction program: Include targets for reduction of leakage. Reporting on leakage and recycling progress will, at least initially, be partly activity based and partly results based. Utilities will need to calculate the true potential in water zones and then assess what is cost-effective to pursue, set a goal, and proceed to achieve this goal.

OVERVIEW OF STATE AND LOCAL WATER CONSERVATION PROGRAMS

There are a variety of local, regional, state, and federal entities interested in water conservation programs proposed and undertaken by local water utilities. In some cases, there are state or regional requirements enacted as laws or regulations that must be followed in the development of a plan and in its implementation. This section discusses federal, state, and local examples.

USEPA Guidelines

The United States Safe Drinking Water Act of 1996 required that the USEPA develop guidelines for water conservation plans to address three categories of public water supply systems serving (1) fewer than 3,300 persons, (2) between 3,300 and 10,000 persons, and (3) more than 10,000 persons. Drinking Water State Revolving Fund programs can require applicants to develop a plan that meets these guidelines as a condition for financial assistance.

These categories are referred to as small, medium-size, and large water utilities. In general, the majority of utilities are very small, serving fewer than 500 persons. In California, for example, the large majority of the state's 2,900 utilities each serve fewer than 500 persons. It has been recommended that all but the very small utilities should have a minimum level of water conservation program. Utilities serving fewer than 500 persons could be exempted from the conservation planning requirement, and small utilities could have reduced requirements. Because of the higher potential for water conservation savings, large utilities should be required to analyze water conservation measures in addition to those in the minimum plan and should submit a formal water conservation plan.

One way that states can handle the utility size issue is to adopt the following guidelines:

- No plan required for utilities serving fewer than 500 persons
- No plan required for systems serving 500 to 3,300 persons unless in an area designated by the state as a "critical water supply area"
- A minimum plan/program for systems serving between 3,300 and 10,000 persons
- A full conservation plan for systems serving more than 10,000 persons

State Requirements

Twenty-eight states have some form of conservation planning or water loss monitoring requirements (the remaining 22 do not as of 2012). California, Arizona, Georgia, Texas, Colorado, Washington, Kansas, and Rhode Island are some examples. In a 2012 study undertaken by the Alliance for Water Efficiency and the Environmental Law Institute (AWE and ELI 2012), all 50 states were evaluated for water conservation requirements that are embedded in laws and regulations (www.allianceforwaterefficiency.org/2012-state-information.aspx). A standard set of 20 questions were asked of every state. The report found that some states have water conservation planning requirements that are tied to the permitting process, other states incorporate planning with funding eligibility as is suggested in the Safe Drinking Water Act, and others require that all public water suppliers submit regularly updated plans.

In another study (Rashid et al. 2010), progress in US water conservation planning and implementation was tracked from 1990 to 2009. It was found that nine states had planning requirements in 1990 and 23 states had requirements by 2009 (Figure 2-4). About 20 other states recommend or encourage conservation. Population growth and water shortages were found to be the main drivers for these 23 states to take a more proactive role in

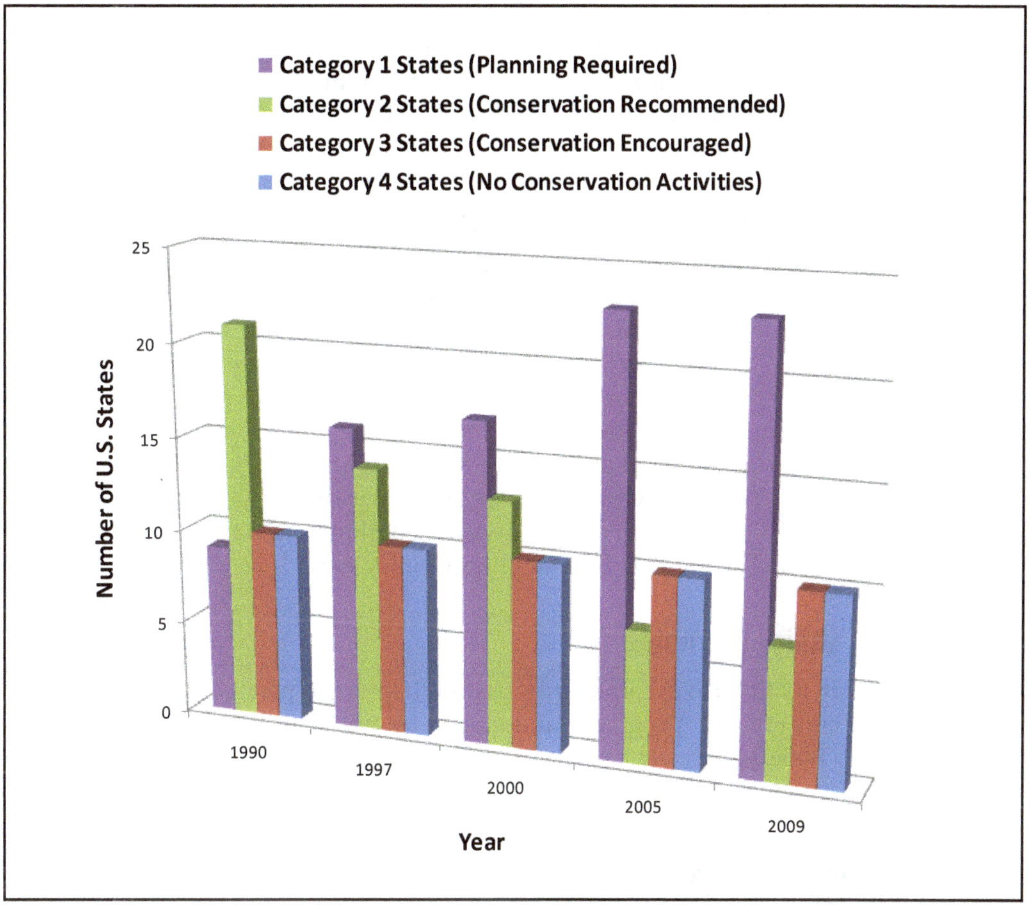

Figure 2-4 Progress of state-level water conservation activities 1990–2009
Source: Rashid et al. 2010

conservation. As these state guidelines and requirements change, it is important for plan preparers to make sure they have the latest standards.

Specific State Examples. The following list contains information on activities that have been undertaken by leading state programs.

- **Arizona:** Water utilities must report their consumption in gallons per capita per day to meet designated targets. See http://www.azwater.gov/AzDWR/WaterManagement/AMAs/PhoenixAMA/RegulatoryPrograms.htm. See sidebar "State of Arizona Municipal Water Use Requirements."
- **California:** (1) California requires five-year urban water management plans (UWMPs) from all water utilities. The law details specific requirements for conservation planning and implementation. See www.water.ca.gov/urbanwatermanagement/. (2) The California Water Efficiency Partnership's MOU specifies best management practices for conservation program implementation. www.cuwcc.org. See sidebar "California's Memorandum of Understanding on Urban Water Conservation." (3) Adopted plumbing standards stricter than those required by the 1992 Federal Energy Policy Act.
- **Colorado:** Adopted plumbing standards stricter than those required by the 1992 Federal Energy Policy Act.

- **Georgia:** (1) The 2010 Georgia Water Stewardship Act requires annual water loss audits of utilities serving 3,000 or more in population. See http://extension.uga.edu/publications/detail.cfm?number=C995. (2) Adopted plumbing standards stricter than those required by the 1992 Federal Energy Policy Act.
- **Kansas:** The Kansas Division of Water Resources requires that utilities prepare water conservation plans to qualify for any state revolving loan fund money or due to a number of other criteria. See http://agriculture.ks.gov/divisions-programs/dwr.
- **Texas:** (1) Texas Council on Environmental Quality requires water conservation plans for certain water rights permits. See www.tceq.texas.gov/permitting/water_rights/conserve.html. See sidebar "Texas Water Conservation Plans." (2) Texas Water Development Board (TWDB) requires water loss audits. All retail public water systems with more than 3,300 connections or a financial obligation to TWDB are required to complete and submit a water loss audit annually. All other retail public water suppliers are required to submit a water loss audit to the agency every five years. See www.twdb.texas.gov/conservation/resources/water-loss-resources.asp. (3) Adopted plumbing standards stricter than those required by the 1992 Federal Energy Policy Act.

Utility Examples

In addition to examples provided in chapters, appendix A includes 13 case studies of water conservation programs undertaken by water utilities in the United States and Canada (Table 2-3).

Table 2-3 Case Studies in Appendix A of AWWA Manual M52

Case Study	Geographic Location	Topic	Supports Chapter(s)
1	Phoenix, Ariz.	State of Arizona Municipal Water Use Requirements	2
2	Tampa Bay, Fla.	Long-Term Supply/Demand Management Plan	3 & 4
3	Philadelphia, Pa.	Water Loss and Water Loss Control in Philadelphia, Pennsylvania	3
4	Abbotsford, B.C.	Using Advanced Metering Infrastructure to Enhance Conservation Efforts	3
5	Seattle, Wash.	Saving Water Partnership – Water Smart Technology Program: Industrial, Commercial, and Institutional Program	3
6	Fort Collins, Colo.	Assessments Lead to Customer Action: Sprinkler System Audit Program – Landscape	3
7	Cobb County, Ga.	Cobb County Water System's Give Them an Inch . . . Grow a Yard Program	3
8	New York, N.Y.	New York City Challenges Hotels to Save Water	3
9	Guelph, Ont.	Graywater Reuse and Rainwater Harvesting Rebate Programs	3
10	Austin, Texas	Mature Conservation Program Shifts Focus From Rebates to Codes Due to Saturation	4
11	Waukesha, Wis.	Comprehensive Municipal Water Conservation Plan for Mid-Size Suburban Community	5
12	Sacramento, Calif.	Engaging Stakeholders in Water Conservation Planning	6
13	Town of Cary, N.C.	Block Leader Program Overview: Environmental Program for Public Education of Water Customers	6

STATE OF ARIZONA MUNICIPAL WATER USE REQUIREMENTS

In the 1970s, the state of Arizona was facing a water crisis. Surface supplies had been fully utilized, and groundwater table levels were falling rapidly because of agricultural and municipal pumping far in excess of the natural recharge rate. Arizona had rights to the Colorado River, shared with California, Utah, Nevada, and Colorado, but the water was hundreds of miles away from the areas in need. The federal government offered to build the canals and pumping stations to bring the water to Phoenix and Tucson, but only if Arizona would agree to reduce groundwater pumping to the safe yield level by 2025. A state law was passed, and it created a new state agency, the Arizona Department of Water Resources (ADWR) (http://www.azwater.gov/AzDWR/WaterManagement/AMAs/PhoenixAMA/RegulatoryPrograms.htm), which devised a plan to reduce groundwater pumping in Arizona's three most populous areas. ADWR decided to reduce extraction in stages and created 10-year plans. The first went from 1980 to 1990; the second from 1990 to 2000; and the third from 2000 to 2010. Water use was slowly reduced, on a per capita basis, and groundwater pumping was reduced as areas switched over to the new (and more expensive) surface water.

ADWR regulates municipal water use in one of three ways. Originally, there was only one program, the total gallons per capita per day program. Now, there are two other programs, the alternative program and the non–per capita program. These programs are defined as follows:

- **Total gallons per capita per day (GPCD) program.** For the first management plan, all water providers were assigned a total system GPCD target based on their 1980 water use, future demand projections, and the potential savings from water conservation measures commonly used in the western United States. The targets were modest and ranged from a 6 percent to an 11 percent reduction over a 10-year period. Generally, the higher the current water use, the higher the target. For the second management period (1990–2000), each provider was assigned a GPCD target based on an extensive analysis of the water use in each provider's area and the conservation potential. A similar approach was used in the third period (2000–2010). Providers who did not meet their targets were subject to fines and sanctions. All but five cities in the state were regulated in this way.
- **Alternative conservation program.** This alternative was created to provide flexibility by establishing a limit on groundwater use, a GPCD target for residential water use only, and requirements addressing specific nonresidential uses. Only one small city has signed up for this program.
- **Non–per capita conservation program.** Factors such as rapid commercial and industrial development and changing population characteristics affected the ability of some providers to meet their GPCD targets. In 1992, the state legislature established this alternative to the GPCD regulatory program. This program requires providers to eliminate groundwater pumping, unless replaced by groundwater recharge, and to implement reasonable conservation measures (RCMs). Four medium-size cities are now using this program (Chandler, Gilbert, Scottsdale, and Tempe). Several others are considering switching to this program, because compliance is generally easier.

Currently, the ADWR specifies 12 RCMs and provides four substitute RCMs that can be used in lieu of the original 12 RCMs. These RCMs cover all aspects of municipal water use from residential to nonresidential. Agencies that adopt this program agree to the water supply requirements, to implementation of the RCMs in a specified way, to an implementation schedule, and to detailed annual progress reporting. Progress reporting covers details on implementation, costs, and estimated water saved per reporting period. Cities can opt for this program at any time and enter into a memorandum of understanding with ADWR to carry out the program. ADWR expects most cities to migrate to the Non–Per Capita Program because of the difficulty of managing total per capita water use. The Non–Per Capita Program is similar to California's use of best management practices (see sidebar "California's Memorandum of Understanding on Urban Water Conservation"), except that compliance is mandatory.

In terms of compliance with management plan requirements, the ADWR can levy civil penalties for use of groundwater beyond the agreed limit. This would be triggered by not meeting the provisions of the selected municipal conservation program, for example, if per capita goals are not met or, depending on the program, agreed-on RCMs were not carried out as specified. The civil penalty can be up to $10,000 per day. In deciding whether to levy a fine or not, the ADWR considers the seriousness of the violation and the extent of the water provider's attempts at compliance and mitigation.

California's Memorandum of Understanding on Urban Water Conservation

The California Water Efficiency Partnership (CWEP) is the only organization operating on a statewide basis that addresses implementation of utility-based water conservation programs. Central to its mission is a Memorandum of Understanding Regarding Urban Water Conservation in California, which was first signed in 1991 and has since been signed by most large water suppliers within the state. The recitals to the MOU present the essence of the agreement and are summarized as follows:

- Signatories recognize the need to provide reliable water supplies and to protect the environment.
- Increasing demands for urban, agricultural, and environmental water uses call for conservation and the elimination of waste as important elements of the overall management of water resources.
- Many organizations and groups in California have an interest in urban water conservation, and the MOU is intended to gain much-needed consensus on a complex issue.
- The urban water conservation practices included in the MOU (referred to as BMPs) are intended to reduce long-term urban demands below what they would otherwise be and are in addition to programs that may be instituted during occasional water supply shortages.
- It is the intent of the MOU for agencies to pursue cost-effective water conservation in a structured manner and consider water conservation on an equal basis with other water management options.
- The signatories agree to use the MOU as the basis for determining future water needs. This is a major step toward achieving consensus on one major issue related to approving new water supply projects. However, the MOU does not provide agreement on other issues surrounding new supplies.

In the 26 years since the original MOU was signed, it has been amended numerous times. The following chart presents a summary of the current best management practices. In 2007, the California legislature passed a law requiring water agencies in California to certify that they were meeting or planning to meet the goals in the BMPs in order to qualify for state grants or loans. In 2008, the CWEP adopted wholesale changes to the BMPs to encourage more aggressive conservation. These changes included giving credit to agencies that met the specific water savings goals with more creative flexible approaches, and those that achieved overall reduction in per capita water demand beyond the level which the original BMPs would achieve.

Urban Water Conservation Best Management Practices for the
California Water Efficiency Partnership

1. Utility Operations
   - 1.1 Conservation coordinator – Designate a water agency staff member to have the responsibility to manage the water conservation program. Water waste prohibition – Adopt water waste rules or ordinances to prohibit gutter flooding, single-pass cooling systems in new connections, nonrecirculating systems in all new car wash and commercial laundry systems, and nonrecycling decorative water fountains. Wholesale agency assistance program – Provide financial incentives to water agencies and cities to encourage implementation of water conservation program.
   - 1.2 Water loss control – Audit the water distribution system annually using the AWWA water loss methodology and proactively control water loss and leaks.
     - 1.2.1 Metering with commodity rates for all new connections and retrofit of existing connections (install meters in 100% of existing unmetered accounts within 10 years; bill by volume of water use; assess feasibility of installing dedicated landscape meters).
     - 1.2.2 Retail conservation pricing – Eliminate nonconserving pricing policies and adopt pricing structure, such as uniform rates or inclining block rates, incentives to customers to reduce average or peak use, and surcharges to encourage conservation. Meet one of two options for volumetric proportionality.
2. Public Education and Information
   - 2.1 School education program – Provide active school education program to educate students about water conservation and efficient water uses.
   - 2.2 Public information program – Provide active public information program in water agencies to promote and educate customers about water conservation.
3. Residential Programs
   - 3.1 Residential assistance/landscape water survey program for single-family and multifamily residential customers – Survey 15% of residential customers within 10 years.
   - 3.2 High-efficiency washing machine rebate program – Provide cost-effective customer incentives such as rebates to encourage purchase of machines that use 40% less water per load.
   - 3.3 Residential WaterSense toilet replacement program – Replace older toilets for residential customers at a rate equal to that of an ordinance requiring retrofit on resale.
   - 3.4 WaterSense toilets in residential new construction – Maintain an incentive program or require WaterSense high-efficiency toilets (HETs) in all new residential construction projects.
4. Conservation program for commercial, industrial, and institutional accounts – Reduce water use by an amount equal to 10% of the baseline use within 10 years through a number of potential business-specific and general measures.
5. Large landscape conservation program and incentives – Prepare water budgets for 90% of all commercial and industrial accounts with dedicated meters; provide irrigation surveys to 15% of mixed-metered customers.

Source: CUWCC, www.cuwcc.org

The following figure shows a Commercial/Industrial Water Audit workshop where attendees are educated on water conservation methods and taught how to conduct audits at their own facilities.

Figure 2-5 Commercial/industrial water audit workshop

Source: Maddaus Water Management

Texas Water Conservation Plans

Texas Administrative Code (TAC) Title 30, Chapter 288, requires utilities across the state to submit water conservation plans to the Texas Commission on Environmental Quality (TCEQ) and/or the Texas Water Development Board (TWDB) no less than every five years. The information reported in these plans assists state agencies in evaluating water usage trends and conservation strategies, projecting the state's long-term water needs, and developing regional and state water plans. Water conservation plan information also allows TCEQ to determine whether requested appropriations of state water by water rights applicants is reasonable and necessary for the proposed use(s) and whether the applicants will conserve and avoid wasting water. In 2011, the 82nd Texas Legislature passed Senate Bill 181, which improved conservation reporting procedures and established a consistent methodology for reporting water use data in gallons per capita per day (GPCD). Along with the conservation plan, utilities must submit an implementation report evaluating progress toward meeting the goals established in the prior planning period.

Submittal triggers: Under state statute and TCEQ rules, the following are required to submit water conservation plans:

- Applicants for a new municipal, industrial, or irrigation water right or an amendment to an existing water right

- Any existing non-irrigation surface water right greater than 1,000 acre-feet/year.
- Any existing irrigation surface water right greater than 10,000 acre-feet/year.

Additionally, TWDB rules require entities applying for Board financial assistance greater than $500,000 and entities with 3,300 connections or greater to also submit water conservation plans.

Plan contents: The water conservation plan establishes a strategy or combination of strategies for reducing the volume of water withdrawn from a water supply source, for reducing the loss or waste of water, for maintaining or improving the efficiency in the use of water, and for increasing the recycling and reuse of water. The utility profile is the foundation for water conservation plan development and ensures that important information and data are considered when establishing conservation targets and goals. It contains information about a utility's water and wastewater systems, as well as service-area population, customer type, connection numbers, water use, projected water demand, and water loss. In addition to the utility profile, water conservation plans must contain at minimum:

- Specific, quantified 5- and 10-year targets for water savings to include goals for water loss programs and goals for municipal use and residential use measured in GPCD;
- Information about the accuracy of the metering devices used to account for the amount of water diverted from the source of supply;
- Universal metering;
- A record management system that includes water deliveries, water sales by customer class, and water losses;
- Measures to determine and control unaccounted-for water;
- A public education program;
- A non-promotional water rate structure;
- A reservoir systems operations plan;
- A means of implementation and enforcement such as an ordinance indicating official adoption of the water conservation plan by the provider; and
- Documentation of coordination with the regional water planning group for the supplier's service area.

Utilities with populations of 5,000 persons or more must also have a program for leak detection and repair, and require new and renewed wholesale water supply contracts to include water conservation measures.

If the requirements above are insufficient for a utility to achieve its conservation goals, they may also select any combination of the following strategies:

- Conservation-oriented water rates
- Plumbing codes for efficient fixtures in new facilities and/or retrofit programs for existing facilities
- Reuse/recycling of treated wastewater and/or gray water
- A program for pressure control
- Ordinances for landscape water management
- Incentives for landscape conversion (Figure 2-6)
- A method for monitoring the effectiveness of the water conservation plan
- Any other conservation practice that is appropriate for achieving the goals of the water conservation plan

TCEQ may require implementation of any of these additional strategies if it determines that they are necessary for a utility to achieve its conservation plan goals.

Rhode Island Water Supply Plans and Conservation Requirements

The Water Supply Management Act (Rhode Island General Law 46-15.4 et seq.) was first enacted in July 1991. A subsequent report, Water Supply Analysis for the State of Rhode Island (A.D. Little 1990), predicted that under a moderate supply and demand and system management scenario that sufficient water would be available for all of the state's needs through year 2010. This led to passage of the Water Supply Management Act in 1991 to ensure the demand and system management elements would in fact be implemented. The Act designates the Department of Environmental Management as the responsible authority to see that the provisions of the Act were carried out and to review mandatory water supply system management plans (which include water conservation) that must be filed by all wholesale or retail water utilities serving 50 million gallons or more per year. Regulations implementing the 1991 Act have been updated regularly, most recently in 2011 (http://sos.ri.gov/documents/archives/regdocs/released/pdf/WRB/6393.pdf). The regulations include per capita targets for water utilities. Guidelines for preparing the plans are located at www.wrb.state.ri.us/work_programs_wssmp.html.

Utilities are required to review and update the plan every five years. Water suppliers must also report progress on plan implementation every two and a half years. These reports include updates on demand and production data.

REFERENCES

Alegre, H., W. Hirner, J. Baptista and R. Parena, 2000. *Performance Indicators for Water Supply Services*. Manual of Best Practice series. London: IWA Publishing.

Alliance for Water Efficiency and Environmental Law Institute (AWE and ELI). 2012. *The Water Efficiency and Conservation State Scorecard: An Assessment of Laws and Policies*. Chicago, IL: AWE and ELI.

American Water Works Association (AWWA). 2007. *Water Resources Planning, 2nd ed.* Manual M50. Denver, CO: AWWA.

AWWA. 2009. *Water Audits and Loss Control Programs*. Manual 36. Denver, CO: AWWA.

AWWA. 2010. AWWA Water Loss Control Committee Free Water Audit Software, version 5.0. https://www.awwa.org/resources-tools/water-knowledge/water-loss-control.aspx.

California Water Efficiency Partnership (CalWEP). https://www.calwep.org/.

Little, A. D. 1990. *Water Supply Analysis for the State of Rhode Island: Final Report to Rhode Island Water Resources Coordinating Council*. Rhode Island. Water Resources Coordinating Council.

Ofwat/Defra. 2012. *Ofwat Review of the Calculation of Sustainable Economic Level of Leakage and Its Integration With Water Resource Management Planning*. Birmingham and London: UK Environment Agency Office of Water Services (Ofwat) and Department of Environment, Food and Rural Affairs (Defra).

Pacific Gas and Electric (PG&E). https://www.pge.com/includes/docs/pdfs/shared/saveenergymoney/rebates/clotheswasher_application.pdf

Plumbing Efficiency Research Coalition (PERC). September 2015, revised March 2016. "The Drainline Transport of Solid Waste in Buildings – Phase 2.0." PERC. http://www.plumbingefficiencyresearchcoalition.org

Rashid, Mamunur; William Maddaus; and Michelle L. Maddaus. 2010. "Progress in US Water Conservation Planning and Implementation – 1990–2009." *Journal AWWA* 102:6.

Southern California Edison (SCE). http://www.allianceforwaterefficiency.org/uploadedFiles/News/NewsArticles/ACEEE-AWE-Report-E131.pdf

Stevens, Jonathan. 1990. Texas Water Development Board (TWDB). 2007. *An Analysis of Water Loss as Reported by Public Water Suppliers in Texas.* Prepared by Alan Plummer Associates and Water Prospecting and Resource Consulting, Fort Worth, Texas. Austin: TWDB.

UK Environment Agency. 2013. Demand Management Bulletin, Series 2, Issue 8. London: UK Environment Agency.

Water Research Centre (WRC). 1994. *Managing Leakage, Reports A-J.* Engineering and Operations Committee. Swindon, UK: WRC.

Water Research Foundation (WRF). 2014. Effective Organization and Component Analysis of Water Utility Leakage Data. Denver, CO: WRF.

This page intentionally blank.

Chapter 3

Analysis of Water Use and Water Savings

It is not possible to determine the benefits of conservation without first estimating the rate of growth in water demand, both with and without conservation. Rates of increased demand can be slowed by conservation, even though demand may increase because of demographic influences. Conservation measures can reduce both the average and the peak demand. Conservation will decrease operations and maintenance (O&M) expense as well as capital expenditures.

A detailed understanding of current water use and an accurate forecast of future water demand are essential for making decisions about the nature and scope of a water conservation program. To estimate water savings of potential conservation measures, the characteristics of water use, such as the seasonal pattern and per capita use values, must be measured or estimated. This chapter provides the following information:

1. Understanding existing customers' water use, including two alternative methods for forecasting future water use
2. Assessing the level of water losses and setting leakage reduction targets to forecast future water losses and non-revenue water
3. Identifying and screening water conservation measures
4. Estimating the water savings from conservation measures

EVALUATE CURRENT WATER USE AND PROJECT FUTURE USE

The purpose of this chapter is to describe the steps and methods to develop a baseline water demand projection, without the effects of water conservation (which will be addressed later in this chapter). The steps include (1) describing the relevant demographic characteristics, (2) describing pertinent water use characteristics, (3) analyzing historical water use, and (4) forecasting future water demand.

Table 3-1 Example of Service Area Characteristics

Service Area Characteristic		Value (Example)
Current population		100,000
Future population	5 years	110,000
	10 years	120,000
	20 years	135,000
Number of residential service connections	Single-family	20,000
	Multifamily	5,000
	Total	25,000
Number of nonresidential connections	Commercial	3,000
	Industrial	500
	Public	500
	Total	4,000
Current employment (jobs)		60,000
Future employment (jobs)	5 years	70,000
	10 years	80,000
	20 years	100,000

Describe the Service Area

Table 3-1 can be used as an example to characterize the service area in ways that are useful for forecasting future water use. The table includes sample numbers to demonstrate methodology. Population projections are usually available from local governments and/or regional planning agencies. Employment projections (jobs, unemployed residents) are usually available from the same regional planning agencies or from transportation planning agencies. A forecast of 20 years is usually adequate for demand management planning because this matches up with long-term water supply planning and provides sufficient time to measure the effects of demand management programs.

Describe Water Use

The following best available data are needed to enable water use data to be properly analyzed:

- Demographic data and population growth projections
- Monthly water production data
- The number of water accounts by customer class
- Monthly water sales (usage) data by customer class

If certain types of data are not available or are inadequate, estimates must be used. Data gaps can be filled by using data from similar water utilities or research projects. Figure 3-1 shows how water is generally used in single-family homes in the United States. Outdoor use (primarily for landscape irrigation) is highly variable and depends on rainfall and temperatures during the growing season.

Table 3-2 is an example worksheet that can be used to characterize existing water use. Sample numbers are shown to demonstrate the methods. Characterization includes the following:

Average annual water supplied is the current amount of raw water withdrawn from a source or purchased from a wholesaler and imported or pumped into the service area. If growth has been minor, the water produced over the past few years can be averaged.

Unmetered water use can be estimated. If from a pumped source use the following:

$$\text{Pumping volume} = \text{pumping rate} \times \text{time of operation} \tag{3-1}$$

Time of operation can be obtained from the electric meter by reading the kilowatt demand of an operating pump over a set time interval and using the electric meter monthly billing reads to calculate estimated hours of pump operation (commonly called run time).

Non-revenue water (NRW) can be calculated by conducting a system water audit. The assessment of water losses is discussed later in this chapter. As used in Table 3-2, the value should be the water that is used but not billed to customers, whether it is authorized and metered or not. Non-revenue water can be relatively low (such as less than 10 percent) in a new, well-managed system, with greater losses found in an older system (such as over 20 percent). Determining the cost of the NRW is a better indicator of efficiency than simply a percentage of water lost. Non-revenue water is now widely accepted as the preferred terminology to use instead of the term *unaccounted-for water*, which is too imprecise and had become a poorly defined catchall term.

Peak-day ratio is water produced on the highest water use day divided by the water use on the average day (annual use in gallons).

Estimated seasonal use is the percentage of water use above some minimum base month of the year indoor water use. Indoor water use is generally the low month's water use prorated over the year. Seasonal use is typically associated with outdoor use, such as landscape watering.

$$\text{Seasonal use, \%} = \frac{100 - 100 \times (\text{lowest month})}{(\text{annual use} - \text{unmetered seasonal use})} \tag{3-2}$$

Average water use by customer class may or may not be available from customer billing records. Depending on the categories used by the utility, complete Table 3-2, expressing the results on a million gallons per day (mgd) basis. In some cases, water use data are only available by meter size. The smallest meters are usually reserved for single-family homes (and some small businesses). Larger meters are for apartment complexes, commercial establishments, schools, and industries. Meter size data can usually be used to break water use down into residential and nonresidential, unless multifamily is the predominant type of residential dwelling. Estimating the seasonal use by customer class helps to target conservation programs aimed at reducing peak use.

In appendix A, the Tampa, Florida case study gives an example of analysis of water use by customer class for the service area. The City of Abbotsford, B.C. case study shows how smart meters can be used to assist profiling customer use characteristics, such as peak use.

Check the reasonableness of the data. The following guidelines can be used for the distribution of water use:

- Indoor per capita residential water use may be about 50 to 70 gallons per capita per day (gcd); single-family use is almost always higher than multifamily use.
- Outdoor per capita residential water use varies from a small value (5–20 gcd) in multifamily buildings to an always larger value in single-family homes (10–80 gcd); the range is due to climate and geography.

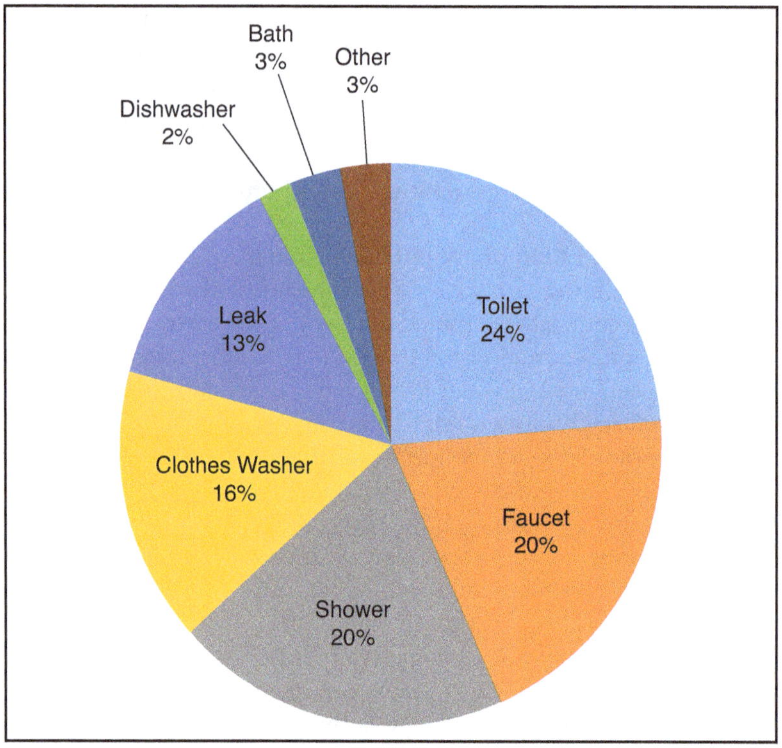

Figure 3-1 Average indoor end uses of water in single-family homes in the United States

Source: WRF 2016

Table 3-2 Example of a Detailed Water Use Description

Water Use Characteristic	Description of Use Type	Value (Example)
Average annual water use (production)		6,570 MG/year (18 mgd)
Non-revenue water		2.7 mgd
Total metered water use		15.3 mgd
Peak-day use		28.8 mgd
Peak-day to average-day ratio		1.6
Estimated metered seasonal use	Month with lowest demand	February
	Average demand in this month	10.7 mgd
	Nonseasonal water use	70%
	Seasonal use	30%
Average metered water use by customer category	Single-family residential	8.4 mgd
	Multifamily residential	1.5 mgd
	Commercial	2.3 mgd
	Industrial	1.6 mgd
	Public	1.5 mgd
	Non-revenue water	2.7 mgd
	Total	18 mgd
System supply safe yield		40 mgd
System capacity		32 mgd

Note: MG/year = million gallons per year; mgd = million gallons per day

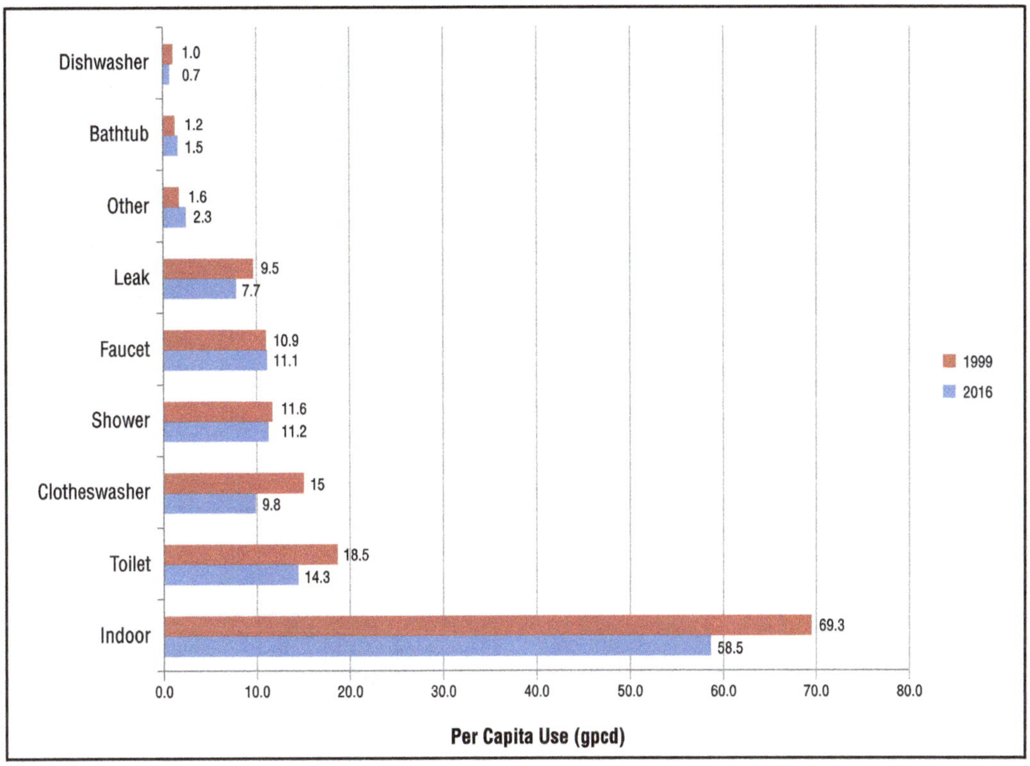

Figure 3-2 Reduction in per capita residential indoor water use 1995–2010

Source: Residential End Uses of Water, 2nd ed., WRF, 2016

- Commercial water use per employee can vary considerably but is often comparable to per capita indoor residential water use, except it is expressed on a gallons per employee per day (ged) basis. It varies with the water intensity of the particular activity at the commercial site.

Trends in indoor per capita residential water use. Figure 3-2 shows per capita water use in single-family homes has changed from 1999 to 2016. Overall indoor use has dropped 11 gcd (16 percent) because of the use of more efficient fixtures, mainly toilets and washing machines.

Analyze Historical Water Use

In addition to completing Table 3-2, it is useful to analyze how water use has fluctuated over the past five or more years. Some of the most common reasons for changes in water use patterns include the following:

- Climate and weather conditions
- Growth (or decline) in water accounts, type of housing, or population served
- Changing demographics of customers (household size, age, income)
- Economic activity and local industrial use
- Price of water and water/sewer rate structures and changes over time
- Water system operation and system losses
- The effect of code changes (toilets, washing machines, landscape and irrigation)
- Long- and short-term conservation actions, including droughts and mandatory restrictions

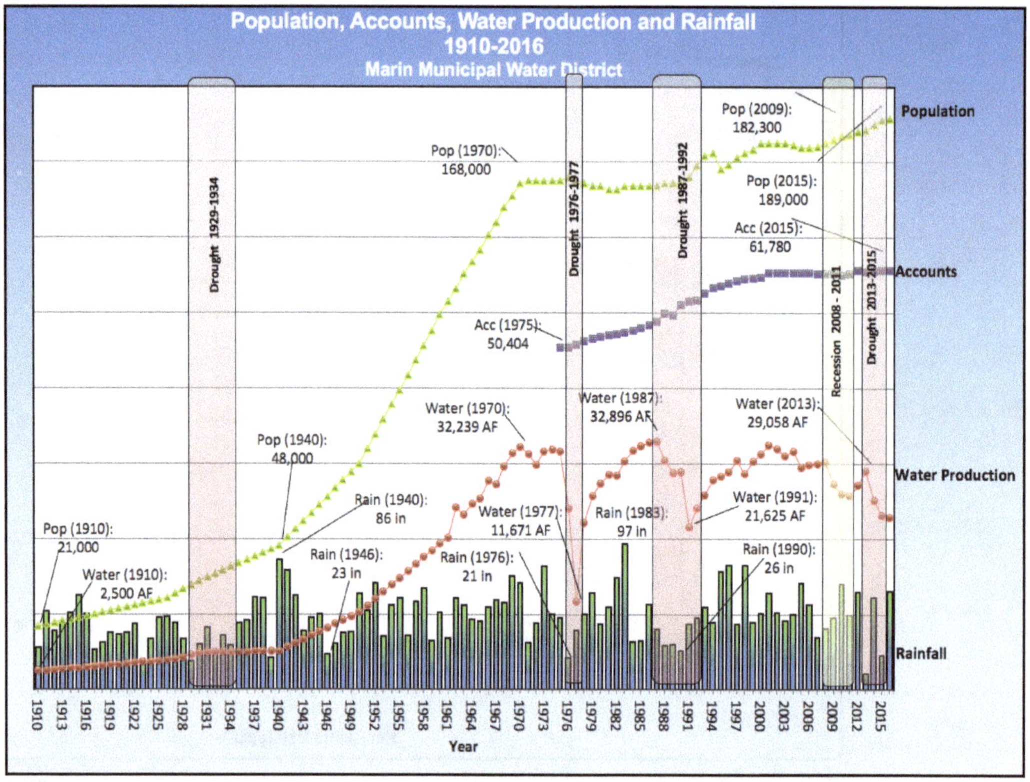

Figure 3-3 Historical annual water use in Marin Municipal Water District (MMWD), California

Source: Lahaye 2016

Figure 3-3 gives an example of water production, water accounts, and rainfall for a long historical record. This type of graph is helpful for studying how a water district has grown over the years and how it responds to increasing population, droughts, and recessions. Note that in this example, water demand was linearly related to population growth until the 1970s. Since then, shifting demographic patterns, conservation, the lingering effects of shortages, and other factors have caused demand to level off and to fluctuate based largely on the weather and local factors. Note how rebounds from droughts never restore water demand to former levels – there are usually some permanent reductions.

If historical monthly water use data are available, a spreadsheet should be developed charting use versus time. To normalize the effects of growth, the total monthly water use should be divided by the number of accounts billed for the month. Figure 3-4 is an example of the seasonal fluctuations for single-family water use within a water service area with a humid climate in the eastern United States. Note how the water use almost doubles between winter and summer. A 12-month moving average shows trends, including three distinct levels of consumption. In addition, over the past 20 years, per-account water use has dropped about 25 percent. In 2001, the impact of water use restrictions to deal with a drought is shown, with another occurring in 2008, after which a recession occurred, keeping the water use depressed. This indicates that using an average or statistical analysis to establish a baseline "normal" water use level is a good idea.

In contrast, Figure 3-5 is an example of single-family water use in a coastal arid area of the western United States. Note how water use can increase by a factor of 5 to 6 from winter to summer. This area also went through a prolonged drought. Water use stayed at a lower level than pre-drought levels because of permanent water use habit and other changes. In the period 2008–2012, the decline was primarily a result of an economic

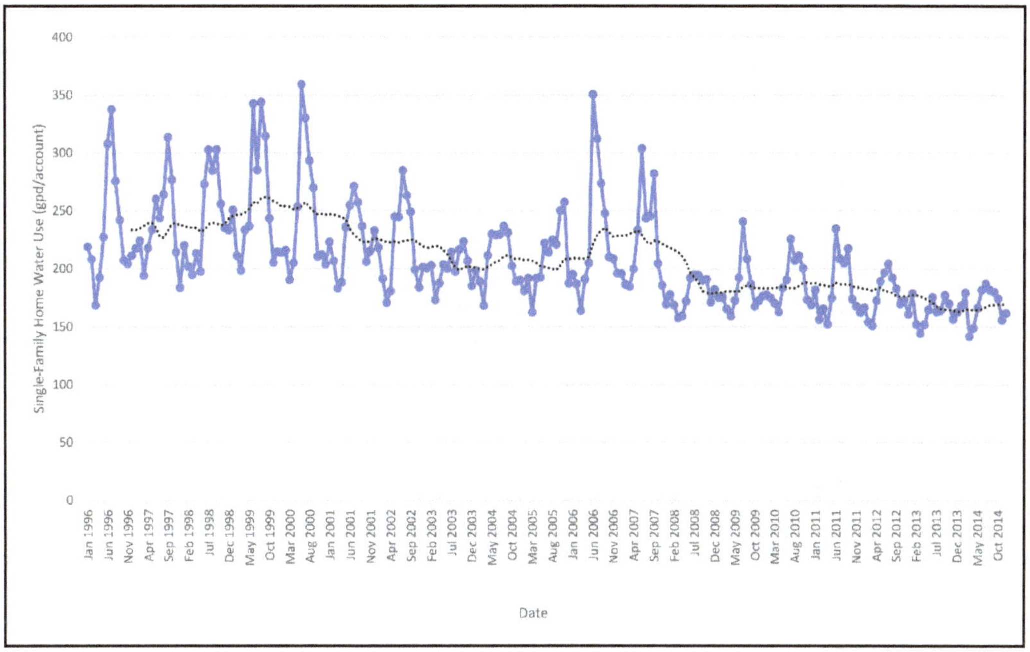

Figure 3-4 Analysis of 20 years of historical single-family residential water use – example of a humid area, Atlanta, Georgia

recession. Again the presence of three base periods of water use in the past 30 years is demonstrated.

Forecast Future Water Demand

In addition to understanding current water use, a detailed forecast of future water needs before conservation is necessary for proper planning and evaluating of conservation benefits. All of the previously mentioned aspects of water use change can influence future water demand.

The subject of demand forecasting is important for a utility, and sophisticated economic and demographic forecasting methods are available and should be considered before a utility makes capital investments to meet future supply. In this manual, two simple approaches to forecasting water use are presented in Table 3-3. The second method is usually adequate for general conservation planning purposes. For the purpose of this illustration, the sample data in Tables 3-1 and 3-2 are used to demonstrate the methods. At a minimum, to evaluate conservation water savings, we need water demands by customer class. These customer-class water demands can be generated using the simple methods presented in this manual. More complex methods of forecasting water demands are presented in AWWA M50 *Water Resources Planning* (AWWA 2007).

Method 1 – Per capita water use: This frequently used, but usually inaccurate, forecasting method assumes that growth in water use (total water production) is proportional to population growth and that per capita use will not change in the future. In making this assumption, this method completely ignores all the other trends noted previously as important determinants of water use (drought and recession impacts, increasing water use efficiency in plumbing fixtures and appliances, a changing climate, changing demographics such as average household size and income, etc.). It directly links future demand to future population as follows:

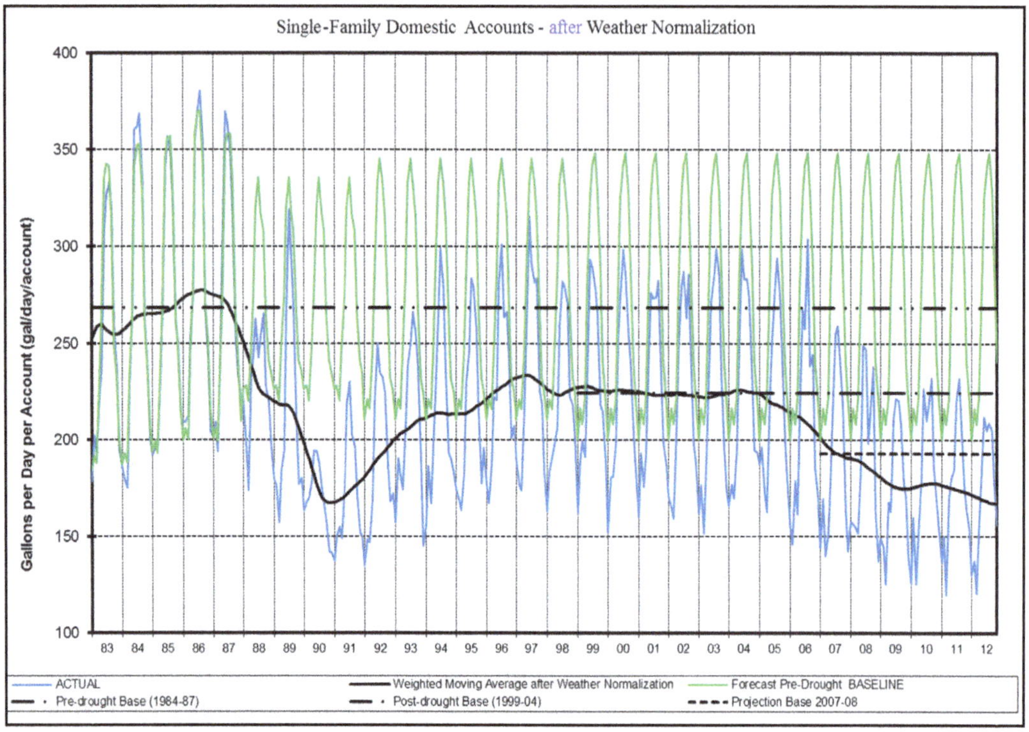

Figure 3-5 Analysis of historical single-family residential water use – example of a coastal, arid area, Santa Cruz, California

$$\text{Future water use} = \text{current per capita water use} \times \text{future operation} \quad (3\text{-}3)$$

Method 2 – Projection by customer class: Figure 3-6 gives an overview of demand forecasting by category of use. This method allows for different growth rates in different water use categories and thus can be more accurate. For example, if employment is growing faster than population, nonresidential water use may grow faster than residential water use. This forecasting method is more sensitive to changes than the per capita water use method.

The second part of Table 3-3 can be used as a worksheet to develop a forecast by customer class. Use factors are developed by residential and nonresidential use. If data are available for additional classes, such as single family and multifamily, further detail can be added.

$$\begin{array}{c}\text{Projected future}\\ \text{water use by class}\end{array} = \text{use factor} \times \begin{array}{c}\text{future population}\\ \text{or employment}\end{array} \quad (3\text{-}4)$$

A community's estimated total water use in future years, or the amount of water needed to be supplied, includes both the water use by customer class and the amount of water lost or unbilled. Water use can be further subdivided into indoor use and outdoor use using the seasonal distribution. Peak-day use can be computed from the overall peak-day factor.

As shown in the example, where employment is growing faster than population, the method 2 projection (rather than method 1) results in a significantly higher projected water use after 20 years, 26.4 mgd versus 24.3 mgd. The difference in peak-day use, 42.2 mgd using method 2 versus 38.9 mgd with method 1, is even more accentuated. Hence, for

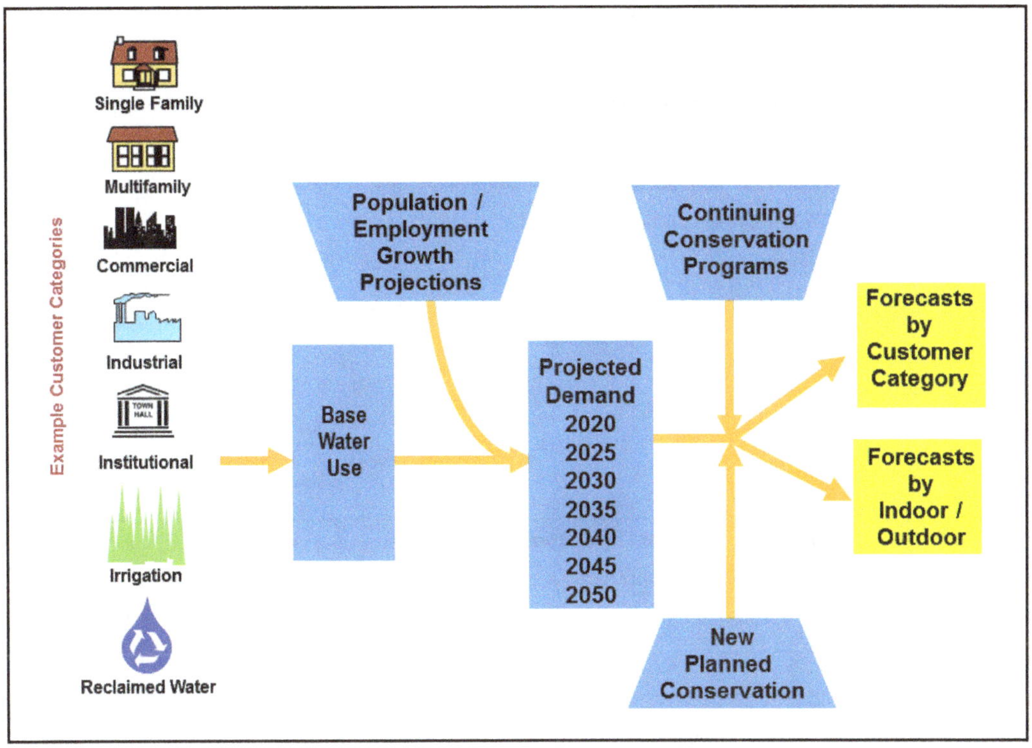

Figure 3-6 Overview of water demand projection process

this example, which anticipates strong job growth, method 2 is preferred and should be used when the data are available.

EVALUATE SUPPLIER ACCOUNTABILITY AND WATER LOSS CONTROL

The actions customers use to conserve water are called *demand-side conservation*. The conservation of water from the supply source to where it enters customer properties is termed *supply-side conservation*, and it is a measurement of water distribution system and management efficiency. In forecasting total water needs, it is important to forecast future changes in the amount of water lost (non-revenue water). Important to supply-side water conservation are two factors: Water utilities must establish accountability in their operations, by compiling the standardized water audit on an annual basis, and water utilities must operate with a reliable degree of efficiency in distributing, metering, and billing the water supplied to customers. These factors have operational, financial, and – most important to conservation – water resources implications. Many water utilities lose millions of gallons of treated drinking water to system leakage, and lose revenue through poor metering and billing practices. Even worse for many utilities, they do not understand the extent of these losses and their cost impacts since they do not routinely audit their supplies. Given in this section is an overview of the methods and tools that are available for reliable water auditing and effective loss control. In addition, Appendix A has one case study relating to water loss control, City of Philadelphia, that exemplifies the material in this section.

Figure 3-7 shows one example of water loss in a system – water loss from main breaks, which is a common occurrence in older systems.

Table 3-3 Example of Forecast Future Water Demand

Method/Parameter	Value (Example)	Value (Example)
Method 1: Per capita water use		
Average annual water use (production)		18 mgd
Current population served		100,000
Current per capita use		180 gcd
Population	5 years	110,000
	10 years	120,000
	20 years	135,000
Future water use	5 years	19.8 mgd
	10 years	21.6 mgd
	20 years	24.3 mgd
Peak-day use – 20 years		38.9 mgd
Method 2: Projection by customer class		
Step 1: Develop unit water use factors		
Residential	Total billed	9.9 mgd
	Population	100,000
	Use factor	99 gcd
Nonresidential use	Total billed	5.4 mgd
	No. employees (jobs)	60,000
Step 2: Project future use		
Residential		
Future population	5 years	110,000
	10 years	120,000
	20 years	135,000
Future residential water use	5 years	10.9 mgd
	10 years	11.9 mgd
	20 years	13.4 mgd
Nonresidential use		
Future employment	5 years	70,000
	10 years	80,000
	20 years	100,000
Future nonresidential use	5 years	6.3 mgd
	10 years	7.2 mgd
	20 years	9.0 mgd
Total future water use	5 years	17.2 mgd
	10 years	19.1 mgd
	20 years	22.4 mgd
Non-revenue water in 20 years		15%
		4.0 mgd
Total future water use — 20 years		26.4 mgd
Peak-day use — 20 years		42.2 mgd

Note: gpcd = gallons per capita per day, ged = gallons per employee per day, mgd = million gallons per day

Figure 3-7 Large water main break

Source: AWWA 2016

Implementing a Reliable Water Auditing Structure

In the United States many versions of a simple water loss percentage indicator were used as the sole measure of water supply efficiency from the 1950s through the early 2000s. The AWWA report *Survey of State Agency Water Loss Reporting Practices* found considerable variation in the percentage definitions employed by state and regional water regulatory agencies (Beecher 2002). Referenced standards for good performance at the time were found to vary from 7.5 percent of total water produced to 25 percent. Since 2003, AWWA has recommended against the use of simple percentage indicators as the sole evaluator of water supply efficiency. In subsequent years, agencies in a number of states and regions have moved away from simple percentage indicators and now require the use of a standardized water audit method and an array of reliable performance indicators (AWWA Water Loss Control Committee 2010).

The International Water Association (IWA), in concert with AWWA, developed a well-defined water audit methodology with a robust assortment of rational performance indicators (Alegre et al. 2000). In 2000 the IWA published this work in the *IWA Manual of Best Practice: Performance Indicators for Water Supply Services* (IWA 2000). The AWWA Water Loss Control Committee published the committee report (2003) "Applying World Wide Best Management Practices in Water Loss Control" in the August 2003 edition of *Journal AWWA* (AWWA 2003). The committee report advocates use of the AWWA water audit methodology and suggests application of innovative loss control methods, which have been applied successfully internationally. The Water Loss Control Committee followed with the development of the AWWA Free Water Audit Software (AWWA 2014) in 2006 and the fully revised version of AWWA's water loss control guidance manual., M36, *Water Audits and Loss Control Programs*, 4th Edition (AWWA 2016). Water Loss Control Committee members have also vigorously addressed this issue to produce a textbook (Thornton et al. 2008) and several research projects on water loss control. Thus, considerable research and development on water loss control has occurred since the early 2000s and many useful

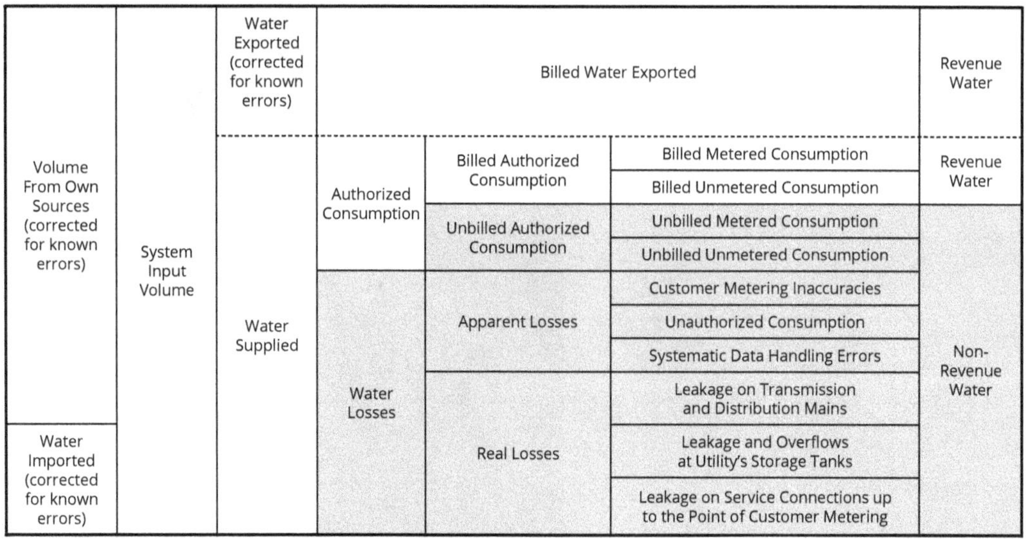

Note: All data in volume for the period of reference, typically one year.

Figure 3-8 Water balance of the AWWA Water Audit Method

references and effective tools now exist to guide and assist water utilities in water auditing and loss control activities.

The AWWA Water Audit Method and Water Balance

A water audit is a compilation of the consumptive uses and losses of the water managed in a single system. The water balance is the summary of water audit components displayed in a format that balances the system input volume with the consumption and losses. The format of the AWWA Water Balance is illustrated in Figure 3-8. The terms in Figure 3-8 are defined in Table 3-4.

Water losses occur in two broad manners. *Apparent losses* are the nonphysical, or "paper," losses including meter and billing system error and unauthorized consumption. Controlling apparent losses is important to conservation programs because the losses often appear as errors in customer billing data and can noticeably skew consumption data needed to evaluate conservation measures. *Real losses* are physical losses, including leakage from distribution mains, customer service lines, and overflows from distribution system tanks or other water storage facilities. Reducing real losses has a direct conservation benefit, as every gallon of reduced leakage is one less gallon that must be pumped from a well, river, or other source, while still meeting legitimate customer demand.

While apparent and real losses differ in the way that they occur, they differ even more dramatically in the impacts that they create. Financially, real losses are often valued at the marginal costs to treat and pump water. Apparent losses are valued at the retail rate that the water supplier charges its customers, a cost usually much higher than the production costs. In cases where water resources are greatly constrained, it is appropriate to value real losses at the customer retail rate, based on the premise that every gallon of leakage saved can go to supply a current or expected customer. As part of the auditing process, costs are assigned to all components of consumption and loss to gauge the financial impact of each. The methodology for compiling the water audit and quantifying apparent and real losses is detailed in AWWA's manual Water Audits and Loss Control Programs (AWWA 2016). Additionally, water utilities can freely download the AWWA Free Water Audit Software, which provides a standard spreadsheet for all of the water audit data inputs to be entered and performance indicators calculated. The software includes built-in

Table 3-4 Water Balance Terms and Definitions From the AWWA Water Audit Method

Water Balance Term	Definition
Volume From Own Sources	The volume of water withdrawn (abstracted) from water resources (rivers, lakes, streams, wells, etc.) controlled by the water utility, and then treated for potable water distribution.
Water Imported	The Water Imported volume is the bulk water purchased to become part of the Water Supplied volume. Typically, this is water purchased from a neighboring water utility or regional water authority.
System Input Volume	The annual volume input to the water supply system. This equals the Volume From Own Sources plus the Water Imported volume.
Water Supplied	The annual volume of treated water delivered to the retail water distribution system. This equals System Input Volume minus the Water Exported volume.
Water Exported	The Water Exported volume is the bulk water conveyed and sold by the water utility to neighboring water systems that exist outside of their service area
Authorized Consumption	The annual volume of metered and/or unmetered water taken by registered customers, the water supplier, and others who are authorized to do so.
Water Losses	The difference between System Input Volume and Authorized Consumption, consisting of Apparent Losses plus Real Losses.
Apparent Losses	Systematic data handling error (in the customer billing process), all types of customer metering inaccuracies, and unauthorized consumption.
Real Losses	The annual volumes lost through all types of leaks, breaks, and overflows on mains, distribution reservoirs, and service connections up to the point of customer metering.
Revenue Water	Those components of System Input Volume that are billed and produce revenue.
Non-Revenue Water	The sum of unbilled Authorized Consumption, Apparent Losses, and Real Losses. Also, this value can be derived by calculating the difference between System Input Volume and billed Authorized Consumption.

checks and guidance to the auditor. It has proven to be very user friendly and effective and has been downloaded thousands of times.

The AWWA Water Audit Methodology Performance Indicators

An advantage of the AWWA Water Audit Method is its array of robust performance indicators. Unlike the simple and inconsistently defined percentage indicators of the past, the AWWA Water Audit Method performance indicators reveal objective financial, operational, and water resources performance standings of water utilities as shown in Table 3-5. Normalized operational performance indicators exist for both apparent losses and real losses in volume per service connection per day, or – for low service density systems for real losses – volume per mile of pipeline per day. These indicators allow for an articulate evaluation of losses and strategic planning for loss control interventions. These performance indicators give water utilities an effective means to track their loss control performance and make comparisons with peer utilities.

While many water utilities now compile an AWWA water audit, relatively few of the audits have undergone a detailed validation process. However, to date, two efforts of note have focused on water audit data collection and validation; the data from these two efforts are currently the most reliable in making performance comparisons among water utilities. These efforts are reported on in the next section.

Table 3-5 AWWA Water Audit Method – Performance Indicators

Function	Performance Indicator	Comments
Financial: Non-revenue water by volume	Volume of non-revenue water as a percentage of System Input Volume	Easily calculated from the water balance, has limited value in high-level, financial terms only; it is misleading to use this as a measure of operational efficiency.
Financial: Non-revenue water by cost	Value of non-revenue water as a percentage of the annual cost of running the system	Incorporates different unit costs for non-revenue components; good financial indicator.
Operational: Apparent Losses (OP23)	[gallons/service connection/day]	Basic but meaningful indicator for apparent losses. Easy to calculate once apparent losses are quantified.
Operational: Real Losses (OP24)	[gallons/service connection/day] or [gallons/mile of mains/day], if service connection density is <32/mile	Best of the simple "traditional" performance indicators. Useful for target setting, limited use for comparisons between systems.
Operational: Real Losses Indicator	[gallons/service connection/day/psi of pressure] or [gallons/mile of mains/day/psi of pressure], if service connection density is <32/mile	Easy to calculate this indicator if the Infrastructure Leakage Index (ILI) is not yet known; useful for comparisons between systems.
Operational: Unavoidable Annual Real Losses (UARL)	UARL (gallons) = (5.41Lm + 0.15Nc + 7.5Lc) × P, where: Lm = length of water mains, miles Nc = number of service connections Lc = total length of private service connection pipe, miles = Nc × average distance from curb stop to customer meter, Lp (Note: The Lp distance is zero if customer meters are located at the property line separating utility/customer ownership.) P = average pressure in the system, psi	A theoretical reference value representing the technical low limit of leakage that could be achieved if all of today's best technology could be successfully applied. A key variable in the calculation of the Infrastructure Leakage Index (ILI). The UARL calculation is not valid for systems with fewer than 3,000 service connections.
Operational: Real Losses	Infrastructure Leakage Index (ILI) = Current Annual Real Losses (CARL)/ UARL (dimensionless) (Note:CARL is the value of Real Losses derived in the water audit.)	Ratio of Current Annual Real Losses (CARL) to Unavoidable Annual Real Losses (UARL); best indicator for comparisons between systems.

Source: AWWA 2016

Compiling Validated Utility Water Audit Data

Growing awareness of the AWWA Water Audit Method and the availability of tools such as the AWWA Free Water Audit Software has moved many water utilities to undertake water auditing for the first time. State and regional regulatory agencies have also taken interest and water auditing requirements now exist in a number of such agencies. Water utilities usually compile their initial water audit in a top-down fashion, by placing readily available data inputs in the water audit. Unfortunately, most utilities are not routinely collecting reliable data for all of the water audit inputs, and must enact new or improved procedures for data collection. It will take the typical water utility from three to five years of more-probing bottom-up investigations to assemble reliable and validated data that is representative of their operations. The AWWA Free Water Audit Software includes a data

Table 3-6 AWWA Water Audit Compiler – Excerpt of Data Sheet Displaying an Excerpt of Data From the 2014 AWWA Water Audit Data Initiative

Include on Chart	Run Compiler / View Options Name of City / Utility:	Customer Metering Inaccuracies	Systematic Data Handling Errors	Systematic Data Handling Errors Default Use	Apparent Losses	Real Losses	Water Losses2	Non Revenue Water	Length of Mains	Number of Active and Inactive Service Connection
Yes	City of Asheville	111.220	11.956	Yes	140.844	1,958.789	2,099.633	2,285.180	1236.5	55,256
Yes	**Augusta Utilities**	**202.735**	**71.603**	**No**	**307.087**	**2,694.886**	**3,001.973**	**3,552.620**	**1213.3**	**72,235**
Yes	Austin Water Utility	828.761	0.001	No	945.924	4,024.607	4,970.531	5,095.921	3707.0	215,960
Yes	**Birmingham Water Works Board**	**557.467**	**0.001**	**No**	**645.000**	**11,242.159**	**11,887.159**	**12,339.569**	**3941.0**	**230,018**
Yes	The City of Calgary	334.291	82.627	Yes	525.552	8,526.084	9,051.636	9,476.994	3072.7	312,075
Yes	**Chesterfield County Rural Water Co., Inc.**	**6.456**	**1.598**	**Yes**	**9.978**	**115.171**	**125.149**	**130.422**	**732.0**	**8,243**
Yes	Greater Cincinnati Water Works	308.039	696.500	No	1,096.716	4,873.730	5,970.446	6,972.146	3135.8	246,044
Yes	**Consolidated Utility District**	**17.943**	**0.300**	**No**	**27.152**	**813.118**	**840.270**	**902.268**	**1301.0**	**50,510**
Yes	City of Cranbrook	0.000	0.000	No	2.798	172.402	175.201	189.193	101.5	6,696
Yes	**Cobb County Water System**	**341.584**	**16.730**	**No**	**404.568**	**1,347.804**	**1,752.372**	**1,764.294**	**3150.0**	**178,130**
Yes	Dalton Utilities	195.846	15.831	Yes	231.343	1,204.651	1,435.995	1,534.328	1251.0	37,023
Yes	**DC Water and Sewer Authority**	**527.700**	**1789.500**	**No**	**2,449.800**	**5,621.951**	**8,071.751**	**8,748.651**	**1350.0**	**134,284**
Yes	Ellijay Gilmore Water & Sewer Authority	11.638	1.000	No	15.169	218.215	233.384	283.102	227.0	5,527
Yes	**Eatonton Putnam Water and Sewer Authority**	**2.281**	**0.511**	**Yes**	**5.792**	**74.506**	**80.298**	**101.609**	**145.0**	**8,350**
Yes	City of Griffin	18.795	1.798	Yes	23.769	510.230	533.999	551.539	212.7	11,733
Yes	**Halifax Regional Water Commission**	**129.981**	**0.264**	**No**	**158.629**	**1,504.514**	**1,663.143**	**1,763.626**	**1017.2**	**85,957**
Yes	Las Vegas Valley Water District	2638.000	100.000	No	2,998.997	3,025.078	6,024.075	6,030.775	4515.0	397,526
Yes	**Louisville Water Company**	**973.100**	**150.000**	**No**	**1,123.200**	**4,123.662**	**5,246.862**	**7,839.099**	**4156.0**	**306,079**
Yes	Macon Water Authority	119.744	6.252	No	132.247	1,551.136	1,683.383	1,779.733	1400.0	65,200
Yes	**Orange County Utilities Department**	**104.165**	**32.920**	**No**	**191.107**	**1,841.418**	**2,032.525**	**2,144.747**	**1745.5**	**90,402**
Yes	Philadelphia Water Department	1490.200	3579.300	No	7,495.000	21,267.500	28,762.500	30,721.500	3178.0	527,205
Yes	**The Region of Peel**	**725.152**	**1.321**	**No**	**855.072**	**4,717.505**	**5,572.577**	**6,079.497**	**2793.9**	**315,617**
Yes	Village of Santa Clara	1.254	0.250	No	1.740	20.613	22.353	24.947	25.0	752
Yes	**South Jordan City**	**63.709**	**9.664**	**Yes**	**84.822**	**289.389**	**374.211**	**714.143**	**333.0**	**19,074**
Yes	City of Wilmington	171.726	500.000	No	701.726	1,832.707	2,534.433	2,631.175	410.0	37,751
Yes	**Water & Wastewater Authority of Wilson County**	**5.228**	**0.020**	**No**	**6.170**	**58.944**	**65.114**	**66.494**	**326.5**	**7,052**
Yes	Washington County Service Authority	14.449	3.485	Yes	24.269	1,047.489	1,071.758	1,139.856	852.5	22,500
Yes	**Cherokee County Water & Sewerage Authority**	**87.701**	**4.162**	**Yes**	**103.643**	**310.021**	**413.664**	**549.551**	**1234.2**	**62,708**

grading capability that allows the utility auditor to grade, or rate, the validity, or degree of trustworthiness, of the data inputs. A composite Data Validity Score based on a 1–100 range is calculated by the software. The software also includes a worksheet that presents the operational criteria that match the levels of data grading for each of the components. By comparing the language of the criteria with the gradings provided by the utility auditor, an external auditor can validate the data. The two notable water audit data collection and validation efforts are the Water Audit Data Initiative (WADI) organized by the AWWA Water Loss Control Committee (WLCC) (WRF 2014) and the effort managed cooperatively by the Georgia Department of Natural Resources – Environmental Protection Division, the Georgia Environmental Finance Authority (GEFA), and the Georgia Association of Water Professionals as required under the Georgia Water Stewardship Act of 2010 (GEFA 2017).

The WLCC organizes a continuing voluntary effort that invites interested water utilities to use the AWWA Free Water Audit Software to compile and share their water audit data with the WLCC, which undertakes a validation process. This process includes one or more validation conference calls of utility staff and members of the WLCC. This effort has resulted in the creation of a validated data set of 20–30 water utilities in each year since 2011. The data and analysis are available at www.awwa.org/resources-tools/water-knowledge/water-loss-control.aspx/. The collected data are stored in the AWWA Compiler Software©, a companion tool to the AWWA Free Water Audit Software. The Compiler easily stores and displays water audit data for multiple water utilities (see Table 3-6) and includes graphing capabilities that effectively display the range of values of water audit parameters among the water utilities in the data set. Examples of several charts measuring the normalized values of apparent and real losses for the 2013 data set are shown in Figures 3-9 and 3-10 respectively.

Water audit data collection in the state of Georgia began in 2011 for large systems and in 2012 for small systems, and the first validated data were published in 2013. The published data included more than 200 water utilities of all sizes. Most notable of the Georgia water audit experience is the strong focus on training of water utility auditors and external review and data validation. With guidance from the Georgia Environmental Finance Authority, state revolving funds were used to fund a Water Loss Technical Assistance effort

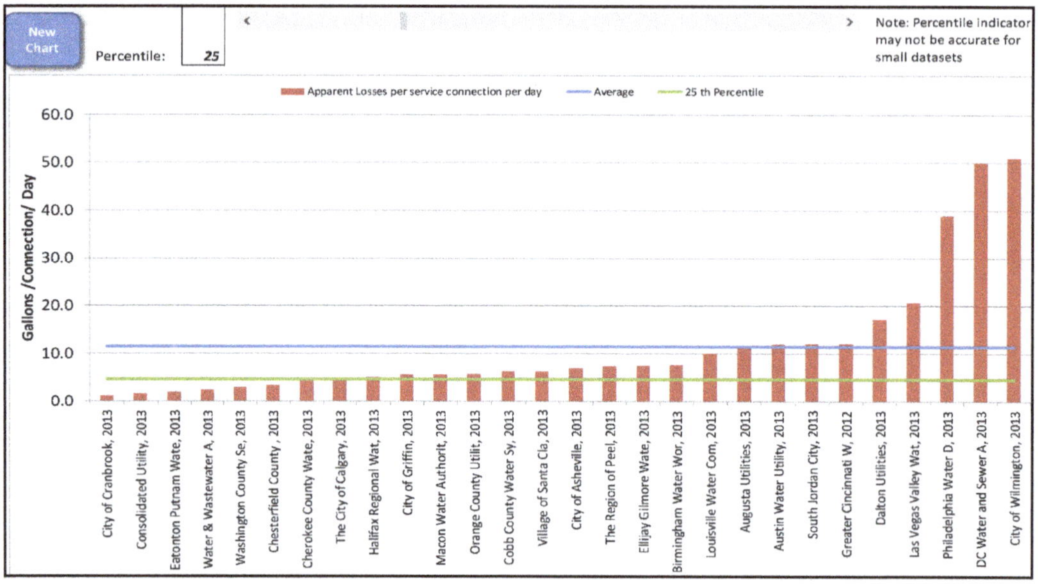

Figure 3-9 AWWA Water Audit Compiler – graph of the apparent losses performance indicator (OP23) for the 2014 AWWA Water Audit Data Initiative

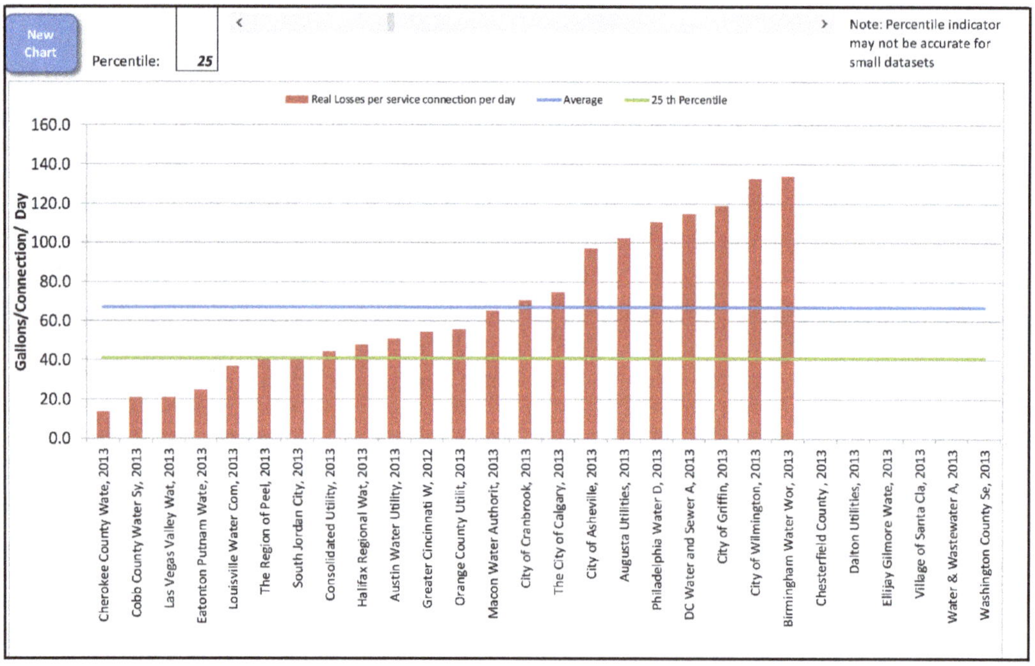

Note: This indicator is not calculated for six water utilities with a low density of customer service connections. Instead an indicator in the form of gallons per mile of pipeline is used.

Figure 3-10 AWWA Water Audit Compiler – graph of the real losses performance indicator (OP24) for the 2014 AWWA Water Audit Data Initiative

that included a number of training sessions held in various locations around the state. Information on this program can be found at http://gefa.georgia.gov/water-loss-technical-assistance (GEFA 2017). Funding was also used to operate teams of trained consultants who validated audit data of the water utilities to verify the utilities' sources of quantitative

Figure 3-11 Accurate water meters are a key to quantifying non-revenue water

Source: Maddaus Water Management

data and that the qualitative data gradings were representative of water utility practices. Water utilities with low data validity scores (50 or lower) should focus on improving their data collection and validity by installing or replacing production and/or customer meters, improving billing system procedures, and beginning investigations into the occurrence of unauthorized consumption. Water utilities with higher data validity scores should continue to seek means to improve their data collection, but can also focus on improving their loss control interventions to control both apparent and real losses. Although Georgia is not the first state to employ the AWWA Water Audit Method, it is the first to develop a comprehensive program that requires water audit data collection and validation, and it provides training and networking opportunities for water utilities to refine their water audit processes and serve as the foundation for strategized water loss control.

Controlling Apparent Losses

Apparent losses do not represent the physical waste of water and – on surface level – may not seem to offer a strong water conservation benefit. However, a closer evaluation of their nature and impact shows that they have a critical role in managing drinking water supplies wisely.

Apparent losses are errors in data and unauthorized consumption. Errors occur because of inaccurate customer meters as well as lapses in estimating, accounting, and billing system processes. A typical water meter is shown in Figure 3-11 and is normally read in gallons or cubic feet. The red triangle indicates flow through the meter and can indicate a leak if water-using activities in the building have ceased. Unauthorized consumption occurs when users bypass or tamper with meters or meter reading equipment to understate the amount of water consumed. Illegal connections and unauthorized withdrawals from fire hydrants are other examples of unauthorized consumption. In all such cases, water suppliers do not accurately account for water consumed by a portion of their customer population. Potential revenue from these customers is lost and customer consumption is understated.

The implications of the apparent losses on water conservation programs include the following:

- Controlling apparent losses recovers uncaptured revenue, which can provide much-needed funding to continue and/or expand the loss control and conservation program.
- Controlling apparent losses reduces the amount of customer consumption error existing in the billing systems that most water utilities use to collect revenue and warehouse customer data. While designed as financial (revenue) management systems, many water utility billing systems are used as de facto customer consumption databases, but often without sufficient controls to ensure that:
  - All water users are documented in established customer accounts (example: many municipal water utilities do not bill or otherwise track water consumption at municipally owned buildings)
  - Accurate meter reading or data on reliable water consumption estimates are entered on a routine basis
  - Inordinate customer consumption patterns are flagged for scrutiny
  - Billing (money) adjustments do not distort actual customer consumption (water) data

To reliably evaluate the success of water conservation programs, customer consumption data must have a high degree of integrity. The water conservation manager should attempt to gauge the apparent losses in a drinking water utility when starting a conservation program. Controlling high apparent losses will help ensure that accurate customer data exist before and after conservation measures are implemented in order to evaluate the impact of the conservation program. Other benefits include the following:

- Controlling apparent losses promotes equity among customer ratepayers because water utilities must effectively spread the cost of scofflaws among the honest ratepayer population.
- Controlling apparent losses projects a strong message to the public that the water utility is closely monitoring water consumption and will act quickly to correct meter or billing errors and thwart illegal activity.

Controlling apparent loss has the advantage of recouping an almost immediate monetary gain in recovered revenue. Water conservation managers are urged to meet with the groups in their utility that handle the metering, meter reading, and billing functions to determine the extent of apparent loss that exists and to gain a better understanding of metering and billing operations in general. Reviewing billing records is a good first step. Matching account inventories to population data or other utility data sets can identify nonaccount water users. Analyzing 12-month billing records can detect unusually high or low (watch for consecutive zero-consumption months) readings. It is likely that a certain amount of apparent loss will be found in any system. Establishing controls to prevent such losses from occurring and programming alerts when they do will help control these losses.

Controlling Real Losses: Leakage and Pressure Management

All water utilities incur leakage; only the type and extent of the leakage varies. All water utilities should therefore maintain some capability to identify and control leakage in their distribution systems, including at meters and service lines. A water audit and leakage component analysis has value in providing information and analysis on the nature, extent,

Figure 3-12 Checking for leaks on the customer side of the meter

Source: Maddaus Water Management

amount, and areas of leakage in a distribution system, providing a good basis to design or optimize the leakage management program. Leakage component analysis is a methodology to collect and categorize data on leakage events occurring in the water utility, and to analyze the leakage patterns in a manner that allows the most cost-effective leakage control interventions to be devised. Figure 3-12 shows a worker listening for a leak at the meter which may also be located on a service line or inside a house. Recent research (WRF 2014) has produced a useful software tool that allows users to develop their leakage tracking and identify the most effective leakage and pressure management strategy for their water distribution system.

Figure 3-13 illustrates the recommended approach to reducing real losses advocated in AWWA Manual M36 (AWWA 2016). See chapter 7 in that manual for a thorough discussion of approaches to leakage and pressure management.

A leak-free water system is not technically feasible or an economical objective, because water is under pressure and some level of leakage will always exist, even in well-run systems. Because the occurrence of leakage is dynamic, all water utilities should exert some effort to combat the emergence of leakage in their system. Unfortunately, most water utilities practice only reactive leakage management – i.e., when a leak or rupture is reported as a visible and/or disruptive incident, utility personnel respond as soon as possible to contain the event and implement a repair. These water utilities are actually only addressing the minority of leakage events occurring in their system, and are allowing a backlog of hidden leakage events to fester. To successfully control leakage, water utilities must practice proactive leakage management by operating programs to identify hidden, or unreported, leakage, which usually constitutes the majority of leakage occurrences (and lost water) in typical water distribution systems.

Proactive leakage management programs include four pillars of intervention (AWWA 2016):

- **Active leakage control:** Identifying and quantifying existing leakage in a distribution system is typically performed using sonic leak detection surveys and continuous monitoring of flows into small zones or district metered areas (DMAs).
- **Speed and quality of repairs:** Optimizing the repair process to ensure timely and lasting repairs is critical to the success of the water loss control program. This is an effort for both the water utility and for utility customers, who usually own part or all of the water service connection piping from the distribution system water mains to their properties, and thus play a role in arranging for repairs of leaks

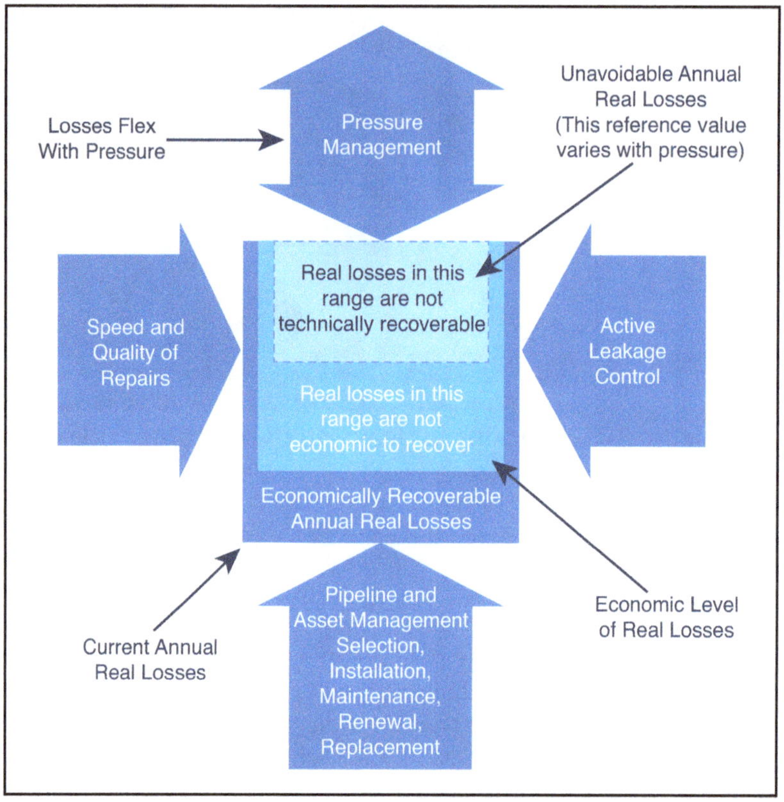

Figure 3-13 The four-pillar approach to the control of real (leakage) losses
Source: AWWA 2016

or failures of their connection piping. (Note: Worldwide, the majority of leakage events and volume of annual leakage losses stem from leaks on customer service line piping rather than distribution system water mains and appurtenances.)

- **Pipeline asset management:** Systems eventually require renewal or rehabilitation. All pipeline assets reach the end of their useful life in due course and must be rehabilitated or replaced if they are to continue to provide service.
- **Pressure management:** Leakage levels can be improved or worsened solely by changes in the level of operating pressure. Background leakage – tiny weeps and seeps in pipe joints and fittings – is very sensitive to the level of water pressure. Pressure management has been found in many cases to be the most effective approach in controlling leakage internationally (GEFA 2017).

Each water distribution system has attributes that are unique for its operation: type, age and condition of piping materials, soil/other sources of corrosion, pressure levels, and other factors combine to influence the nature and extent of leakage occurring in the system. The challenge to utility managers is to find the right balance of interventions to apply in their system to control leakage losses to acceptable levels. Without leakage and pressure management controls (reactive leakage management), water utilities will incur ever-rising levels of leakage and wasted water resources. By moving to a proactive approach to leakage management, water utilities can contain leakage to economic levels, thus optimizing the water delivery process and maintaining a reliable water supply to their customers. Using sonic leak detection equipment is a useful method of identifying and containing leaks in a distribution system (see Figure 3-14).

Figure 3-14 Using sonic leak detection equipment to find leaks in distribution system

Source: Maddaus Water Management IWA Manual

Conclusions on Water Auditing and Water Loss Control for Supply-Side Water Conservation

Controlling losses (apparent and real) is an endeavor of diminishing returns. Losses can never be completely eliminated. Concepts for water loss control consider the economic level of loss control, or the level where the cost of control interventions equals the value of the water saved by those interventions. Methods are continuing to be developed and refined for water utilities to determine the economic levels of loss control for both apparent and real loss. These determinations take into account both the technical costs of the proposed interventions and the costs of water saved, for both apparent and real losses. This approach gives water utilities a rational and defensible basis to establish and fund loss control efforts, and provides motivation for utility managers to move forward in a proactive manner regarding loss control.

The reader is directed to the AWWA Manual M36 (AWWA 2016) and related materials referenced herein for detailed guidance and assistance in implementing the water loss control program for effective supply-side water conservation

IDENTIFY AND SCREEN WATER CONSERVATION MEASURES

The first question to ask is where to start in identifying conservation measures that may be appropriate for the service area? In the beginning, programs usually start simply with plans the utility knows how to implement. As water conservation programs mature, generally more sophisticated measures are considered. Figure 3-15 shows an example of a pathway from the basic to more innovative measures.

It is important for water conservation planners to have fundamental knowledge of water saving devices and measures. There are two good sources of information for this: (1) water conservation plans of water suppliers that have an exemplary conservation

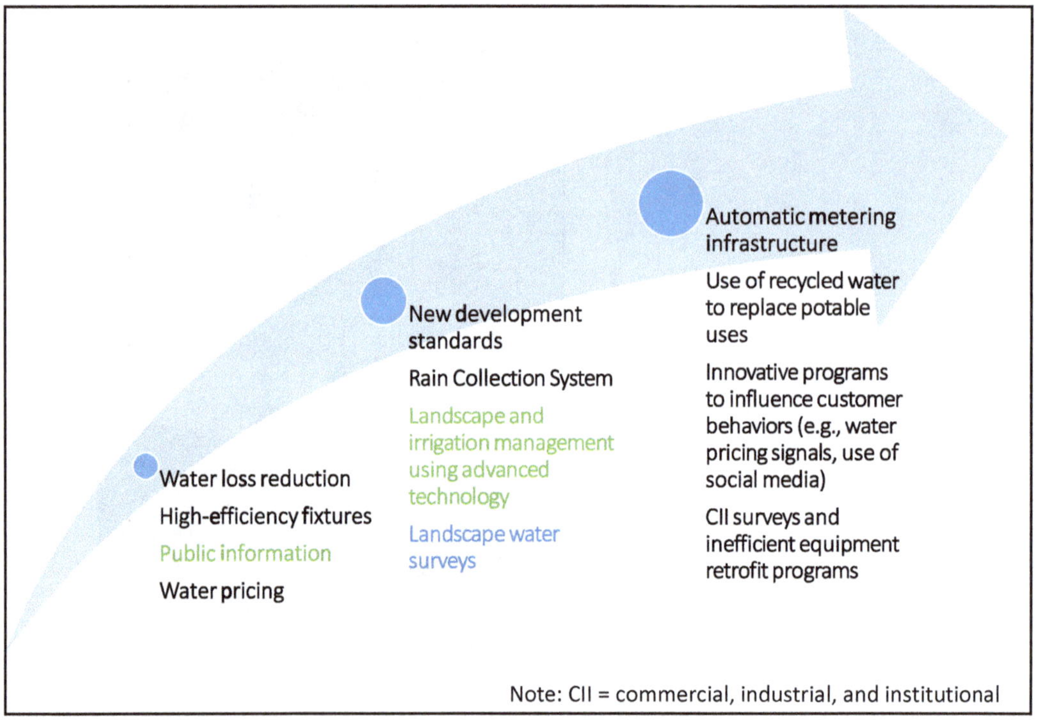

Figure 3-15 Basic to innovative measures typically evaluated

programs, and (2) Internet sites of water utilities that are active in the conservation field. Guidebooks and manuals that have been written on the topic can also serve as a resource for those new to the field. Some are available from AWWA; others can be found on the Internet. Many new professionals will begin their study by accessing information from local and regional information networks. Furthermore, extensive planning guidance is available at the Alliance for Water Efficiency web site www.a4we.org.

Among the case studies in appendix A are four US utilities (Seattle, Washington; Fort Collins, Colorado; Cobb County, Georgia; and New York City) and one in Canada (Guelph, Ontario) that are good examples of specific conservation measures. Other utilities implementing conservation programs are a good source of information on measures that work.

A three-step process is normally used to identify appropriate measures worthy of evaluation for possible inclusion in a conservation program.

1. Identify technologies and devices that could deliver water savings in the service area if installed and used properly.
2. Prepare a list of conservation measures that encourage water efficient technology and practices to utility customers.
3. Screen the measures into a short list worthy of further evaluation.

Efficient Technologies for Customer End-Use Reductions for Fixtures and Fittings

The first step is to compile information on water efficient fittings, fixtures, and appliances into a list of potential measures. Key information to be gathered includes costs, water savings, current requirements of the device, and life of the device. There are various sources of this information; one excellent source is the *Handbook of Water Use and Conservation* (Vickers

2001). This book has extensive coverage of residential fixtures and appliances; landscape water conservation measures; and industrial, commercial, and institutional conservation measures. An update is planned; for details check this web site: www.waterplowpress.com. Another source is the AWWA publication "Guidebook for Small and Medium Sized Utilities" (Green and Maddaus 2010). A new version of this publication is being created and should be published with additional updated information. This provides examples of efficient fixtures and practices. USEPA's WaterSense (www.epa.gov/watersense) and Energy Star (www.energystar.gov) web sites are also excellent sources of water use information on residential and commercial fixtures.

This manual focuses on conservation program planning and design and does not recommend consideration of specific conservation measures. However, a summary of some of the commonly used measures is provided as an example of how to identify and screen them. The devices in the tables are not intended and should not be used as a complete-measure checklist for utilities.

Table 3-7 provides a summary of selected water efficient devices. Actual savings vary with household size, current devices or technology in use, portion of water used in the garden, and so on. Figures 3-16 through 3-18 illustrate cooling towers, usually a major source of water use where employed for cooling large spaces.

Prepare List of Potential Conservation Measures

As part of the evaluation of appropriate measures, a list of potential measures that may be appropriate for the area should be compiled. The list can include existing or ongoing measures currently implemented by the water utility. This process generally yields more than 100 potential conservation measures in the following categories:

- System
  - Water loss, see previous section
  - Automatic metering infrastructure (AMI)
- Residential
  - Single family
  - Multifamily
- Commercial
- Industrial
- Institutional and public or municipal
- Landscape irrigation

Sources of information on conservation measures include the *Handbook of Water Use and Conservation* (Vickers 2001); AWWA "Guidebook for Small and Medium Sized Utilities" (Green and Maddaus 2010); California Water Efficiency Partnership's memorandum of understanding that contains a list of best management practices described in chapter 2; the Arizona Department of Water Resources list of Reasonable Conservation Measures described in chapter 2; and various other states that have conservation guidelines.

A typical list of potential measures is provided in Table 3-8. This table should include all measures being considered. Some of the measures overlap in water savings because they target the same areas for water conservation. This potential overlap can be accounted for, where necessary, during the combination of measures into alternative programs. Often the list contains 50–100 measures; the table is just an example.

Table 3-7 Example List of Water Efficient Devices by Category[a]

		Estimated Fixture Cost (US$)				
Device	Flow Rating[b]	Purchase Cost ($)	Installation Cost ($)	Additional Annual Cost[b] ($)	Device Life (years)[b]	Demand Reduction (gal/conn/day)[c]
RESIDENTIAL						
Bathroom						
Ultra-low-flow showerhead	1.5 ga//min	10–50			5–10	20[d]
High-efficiency showerhead	2 gal/min	10–50			5–10	12[d]
Standard showerhead	2.5 gal/min	10–50			5–10	4[d]
High-efficiency faucet	1.5 gal/min	50–200	50		15–20	12[e]
High-efficiency faucet aerator	1.2–1.5 gal/min	5–10			5	12–16[e]
Standard faucet aerator	2.2 gal/min	5–10			5	4[e]
Toilets						
Ultra-high-efficiency toilet	0.8 gal/flush	90–300	200		30–40	34[f]
High-efficiency toilet	1.28 gal/flush	90–600	200		30–40	28[f]
Standard toilet	1.6 gal/flush	90–300	200		30–40	24[f]
Dual-flush toilet	1.6/0.8 gal/flush	90–400	200		30–40	29[f]
Water displacement device	0.5 gal/flush	5			5	6[f]
Composting toilets	0 gal/flush	1,000–2,000	500	200	20+	44[f]
Kitchen						
Efficient kitchen faucet	1.8 gal/min	90–300	50		10–15	9[e]
Faucet aerator	2 gal/min	5–10			5	6[e]
Dishwasher (domestic)	5 gal/wash	400–900	200		10–15	5[g]
Efficient dishwasher	≤4.25 gal/wash	800–1,400	200		10–15	7[g]
Laundry						
Faucet aerators	2 gal/min	5–10			5	6[e]
Efficient washing machines	14 gal/wash	700–1,000	100		10–15	8[h]
General Household						
On-demand or point-of-use hot water recirculation systems on shower and kitchen sink	2.5 gal/min	500–1,000	200–400		20+	8
Household pressure reducing device	80 psi	50	200		20+	3
Graywater systems		>3,000	>400		15–25	20
Landscaping						
Drip systems		50–100	10–50		10	5
Micro-spray systems		50–100			10	3
Faucet timers		20–50	0–100		5–10	5
Rainwater tanks (150 gal)		1,000	4,000		20+	Varies
Trigger shutoff valves on hoses		10–15			5	1
COMMERCIAL						
General Commercial Equipment (Othan Than Previous Measures)						
Waterless urinals	0 gal	400–500	300	150–200[i]	20+	200[j,k]
Ultra-high-efficiency urinal	0.25 gal/flush	250–800	200–400		20+	175[j,k]
High-efficiency urinal	0.5 gal/flush	250–800	200–400		20+	150[j,k]
Standard urinal	1.0 gal/flush	250–800	200–400		20+	100[j,k]
High-efficiency flush valve toilet	1.28 gal/flush	200–500	200–400		20+	220[j,k]
Standard flush valve toilet	1.6 gal/flush	200–500	200–400		20+	190[j,k]
Ultra-low-flow faucet	0.5 gal/min	50–150	100–200		10	50[l]
Low-flow faucet aerator	1.0 gal/min	5–10	–		5	37[l]

[a] Source: Maddaus Water Management.
[b] Where applicable.
[c] Demand reduction is given in units gallons per connection per day (gal/conn/day). Assume 2.5 persons per connection.
[d] Based on comparison with 2.75 gal/min shower, 8 min duration, 0.8 showers/person/day.
[e] Based on comparison with 2.5 gal/min faucet, 5 min use/person/day.
[f] Based on comparison with 3.5 gal/flush, 5 flushes/person/day.
[g] Based on 10 gal/load conventional and one/load/day/conn.
[h] Energy Star web site assuming 14 gal/load vs. 23 gal/load for conventional and one load/day/conn.
[i] Based on 3–4 cartridges/year (at US$ 50/cartridge).
[j] Per device.
[k] Replacing 2 gal urinals & 3.5 gal toilets, assume10 flushes/day, 10 fixtures in building.
[l] Replacing 2.5 gal/min faucets, assume10 uses/day, 15 sec/use, 10 fixtures in building.

Figure 3-16 Cooling towers are common with large buildings, such as on college campuses

Source: Maddaus Water Management

Figure 3-17 Cooling tower water audits

Source: Maddaus Water Management

Figure 3-18 Cooling tower audits involve measuring conductivity to calculate the cycles of concentration

Source: Maddaus Water Management

Screening of Conservation Measures

Screening Process

The following terms are used in the screening process:

Device. A physical item of hardware, such as a new toilet, or specific action by individuals, such as commercial audits, that would save water if the recommendations are implemented or carried out by the water utility or some other group.

Measure. Device(s) plus a distribution method and possibly an incentive, such as a rebate. A measure can be targeted at a particular type of end user. For example, residential audits would save water if the recommendations are implemented or carried out by the water utility or the resident.

Program. A set of one or more measures targeted at one or more customer classes that would be managed by the water utility as a separate project.

Plan. A set of one or more programs together with an estimated budget, schedule, and staffing plan.

Screening Criteria

Cost-effectiveness. In some states or regions, a list of which measures are considered cost-effective for most utilities may already be compiled. For the purposes of a first screen of measures, only those that are more than ten times the cost of a utility's alternative next source of supply should be eliminated. As a general guide, the measures other utilities believe are effective should be used, or the list in Table 3-8 can be used. Chapter 4 covers evaluation of detailed costs and benefits. If information is available on what the next least costly source of alternative water supply is for the utility, this should be used to avoid spending a lot of time on measures costing ten times or greater.

Technology/market maturity. This screening criterion indicates whether the necessary technology is available commercially and supported by the local service industry. For example, a device may be screened out if it is not yet commercially available in the region.

Service area match. This screening criterion seeks to distinguish the technology that is appropriate for the area's climate, building stock, or lifestyle. For example, low water use landscape measures for commercial sites may not be appropriate where water use analysis indicates there is little outdoor irrigation.

Table 3-8 Example List of Potential Conservation Measures From Southern California

Measure No.	Measure Name	Description
System		
1	Water loss control program: Compute infrastructure leakage index on an annual basis	Conduct an annual audit and adaptively manage for addressing water loss issues. Goal would be to lower the infrastructure leakage index (ILI) and non-revenue water every year by a predetermined amount based on cost-effectiveness. These programs typically pay for themselves based on savings in operational costs (and saved rate revenue can be directed more to system repairs/replacement and other costs).
2	New development: Install advanced metering infrastructure	Require that new customers install smart meters (advanced metering infrastructure or AMI) capable of providing hourly consumption data back to the city and purchase means of viewing daily consumption inside their home/business through the Internet. The AMI system would, on demand, indicate to the customer and city where and how their water is used. thereby facilitating water use reduction and helping customers identify leaks. This system would require the city to fully install an AMI system and then use the system to notify customers of a potential leak (based on continuous flow).
Residential		
3	Public information program including school grants	Public education would be used to raise awareness of other conservation measures available to customers. Programs could include school programs, poster contests, speakers to community groups, media relations, printed educational material such as bill inserts, etc.
4	Conservation pricing	Normally targeted at single-family customers, the rate schedule may be changed from a uniform pricing structure to a three- or four-tiered system where consumption in the upper tiers is charged at a higher unit price ($/1,000 gal) than consumption levels in the lowest tier. A rate study is required to determine prices in each tier and consumption block. To estimate potential water savings, the price elasticity for the service area can be derived from service area data or estimated from relevant published values. For more information about conservation pricing, see chapter 7 in this manual and AWWA Manual M1, *Principles of Water Rates, Fees, and Charges* (AWWA 2017), which has a list of 10 best practices for rate setting.
5	Residential high-efficiency-toilet rebates	Provide a rebate for the installation of a high-efficiency toilet (HET). HETs are defined as any toilet flushing at 1.28 gallons per flush (gpf) or less and include dual-flush technology. Rebate amounts would reflect the incremental purchase cost and have been about $100 per device.
6	Single-family high-efficiency clothes washer rebates	Provide a rebate for the installation of a high-efficiency clothes washer (HECW). HECWs are defined as any washer with a water factor of 4.0 or less and include front loader technology.
7	Plumber-initiated high-efficiency-toilet and/or urinal retrofit program	The city would subsidize installation cost of a new HET purchased in bulk by the city. Licensed plumbers prequalified by the city would solicit customers directly. Customers would get a new HET installed at a discounted price. Up to $1,200 per business, or an average of four replacements. See Figure 3-19.
8	Single-family outdoor water surveys	Conventional outdoor water surveys for existing single-family residential customers. Homeowners with high water use are targeted and provided a customized report on how to save water in their home. Hose nozzles and 5-minute shower timers, aerators, and showerheads provided upon request.
9	Turf removal, single family	Provide $1.00 per square foot incentive (up to $1,000) to remove turf and replace low water use plants or hardscape. For another example program of this type, see the Southern Nevada Water Authority.
10	Mobile home park submetering	Provide a partial cost rebate to meter all sites within a mobile home park that are currently master metered. Pattern after Santa Clara Valley Water District program.
11	New development: Require multifamily individual unit metering on new accounts	Require the metering of individual units in new multifamily units, condos, townhouses, mobile home parks, and business centers (fewer than four stories and with water heater in the units).
12	Existing multifamily individual unit metering incentive	Provide a rebate (per unit) to assist multifamily building owners install submeters on each new individual apartment unit.
13	Residential efficient outdoor use education and training programs	Organize and sponsor a series of educational workshops or other means for educating homeowners in efficient landscaping and irrigation principles. Use guest speakers, native demonstration gardens, and incentives such as a nursery plant coupon. Consider a "river friendly" program that also benefits stormwater quality programs and perhaps co-fund. Existing program example: www.riverfriendly.org.

Table continues next page

Table 3-8 Example List of Potential Conservation Measures From Southern California (continued)

Measure No.	Measure Name	Description
Commercial		
14	Top CII users survey program	All commercial, industrial, and institutional (CII) customers would be offered a free water survey that would evaluate ways for the business to save water and money. The CII surveys would be for large accounts (accounts that use more than 5,000 gallons of water per day) such as hotels, restaurants, stores, and schools. Emphasis will be on supporting the top 100 users. Rebates for equipment replacement would be covered in another measure.
15	CII rebates to replace inefficient equipment	Provide up to a $1,000 rebate for a standard list of water efficient equipment. Indoor equipment included would be high-efficiency toilets and urinals, air-cooled ice machines, connectionless food steamers, cooling tower conductivity and pH controllers, and dry vacuum pumps. Landscape irrigation equipment includes rotating nozzles, large rotary nozzles, and irrigation controllers. Rebate amounts vary. Use Metropolitan Water District's (MWD's) Western Municipal Water District (WMWD) SoCal Water Smart program when funding is available.
16	Customized CII incentive program	After the free water use survey has been completed at site, the city will analyze the recommendations and determine if the site qualifies for a financial incentive. Financial incentives will be provided after analyzing the benefit–cost ratio of each proposed project. Incentives are tailored to each individual site as each site has varying water savings potential. Incentives will be granted at the sole discretion of the city while funding lasts. Program designed to not overlap SoCal Water Smart and cover other items not eligible.
17	New development: Require plan review for new CII	Require plan reviews for water use efficiency for all new business customers. Use incentives to move beyond CALGreen and encourage participation in LEED. Encourage participation in SoCal Water Smart.
18	CII irrigation water surveys	All public and private irrigators of large landscapes (>3 acres) would be eligible for free landscape water surveys upon request. Normally those with high water use would be targeted and provided a customized report. See Figure 3-20.
19	Commercial efficient outdoor use education and training programs	Organize and sponsor a series of educational workshops or other means for educating landscapers and contractors in efficient landscaping and irrigation principals. Use guest speakers, native demonstration gardens, and incentives such as a nursery plant coupon. Consider a "River-Friendly" or "California-Friendly" program that also benefits storm water quality programs and perhaps co-fund. Existing program example: www.riverfriendly.org or www.landscapecontest.com.
20	Turf removal CII/multifamily	Provide $1.00 per square foot incentive (up to $10,000) to remove turf and replace low water use plants or hardscape. Minimum 6,000 square feet, maximum rebate $10,000. For another example program of this type, see the Southern Nevada Water Authority.

Customer acceptance/equity. Customers must be willing to implement measures or else the market penetration rates (and thus the water savings) would be too low to be significant. Customer acceptance may be based on

- Convenience
- Economics
- Perceived fairness
- Aesthetics
- Environmental values

Measures should also be equitable in the sense that one category of customers should not benefit while another pays the costs without receiving benefits.

Screening Process

The measures can be rated in a table such as shown in Table 3-9. As shown in Table 3-9, each measure has been scored on a scale of 1 to 5, with 5 being a high rating. Generally, measures were eliminated that scored mostly 1s and 2s with a few 5s. The screening is qualitative and subjective and can be done by a group that is likely to interpret and score measures differently. The goal should be to reduce the list to about 20 to 30 measures that pass the screen – i.e., have relatively high scores. In general, a measure has to have a score of 12 or more points to pass this screen when the maximum score is 15.

Figure 3-19 illustrates one water saving measure that can be implemented to increase water conservation. Figure 3-20 shows the process of testing root depth with a soil probe, which is a key part of evaluating the efficiency of landscape irrigation (deeper roots are preferred and come with less frequent, longer watering).

EVALUATE WATER SAVINGS

Estimated water savings are useful to help utility planners forecast how future demand may be reduced by water conservation. Savings normally develop at a measured and predetermined pace, reaching full maturity after full market penetration is achieved. This may occur three to ten years after the start of implementation, depending on the implementation schedule.

Methodology

The basic method for estimating water savings is typically expressed as follows:

$$\text{Water savings, } gpd = \begin{array}{c}\text{No. accounts targeted} \times \text{market penetration (\%)} \\ \times \text{ unit water savings, } gpad\end{array} \tag{3-5}$$

There are a number of ways to express the unit water savings. These include gallons per fixture replaced and percent of end use reduced. Equation 3-5 can be adapted to suit available water savings data.

Sources of Data

Data necessary to forecast water savings of measures include baseline water use, end uses of water, demographics, market penetration, and unit water savings. These are described as follows:

Baseline Water Use

Baseline water use (without conservation) projections are usually developed for 20 to 30 years into the future. It may include water losses or just be based on water sales, depending on the needs of the water planner. The baseline water use includes the effects of the current plumbing and appliance codes. The savings caused by the effects of plumbing code can be estimated given assumptions about unit water savings and rate of natural replacement of old fixtures with new, efficient fixtures.

End Water Uses

The base water use or water sales by customer class can be portioned into end uses using the seasonal pattern and research such as the residential end uses of water (DeOreo et al. 2016). An example of end use for a single family is shown along with a customer class breakdown and a typical indoor/outdoor water use split in Figure 3-21. Recent research indicates the average indoor single-family water use has fallen to 58.6 gallons per capita per day, a decrease of 15 percent from 1999 to 2016 (DeOreo et al. 2016).

Unit Water Savings

Units can be individual product units (such as toilets) or groups of products (such as household retrofits), as long as the analysis is consistent. A unit estimate may not be appropriate for each measure, in which case total program water savings and costs for the measure can be used.

Demographics

Service area population, total dwelling units, and residential and nonresidential demand are normally used to evaluate measures.

Table 3-9 Example of Screening Potential Conservation Measures and Results of Screening

Measure			Criteria			Pass
Measure No.	Name	Distribution Method & Incentive	Technology Market Maturity	Service Area Match	Customer Acceptance/ Equity	Score Yes or No
SYSTEM						
1	Water loss control program: Compute infrastructure leakage index (ILI) on an annual basis	Water retailers' sponsor	5	5	5	15 Yes
2	New development: Install AMI	Water retailers' requirement	5	3	2	10 No
RESIDENTIAL						
3	Public information program including school grants	Water retailers' sponsor	5	5	4	14 Yes
4	Conservation pricing	Water retailers' new rate schedule	5	5	2	12 Yes
5	Residential high-efficiency toilet (HET) rebates	Water retailers' rebate	5	5	4	14 Yes
6	Single-family high-efficiency washer (HECW) Rebates	Water retailers' rebate	5	5	4	14 Yes
7	Plumber-initiated high efficiency toilet and/or urinal retrofit program	Water retailers' rebate	3	4	3	10 No
8	Single-family outdoor water surveys	Water retailers' sponsor	5	4	3	12 Yes
9	Turf removal, single family	Water retailers' rebate	5	5	3	13 Yes
10	Mobile home park submetering	Water retailers' requirement	5	4	3	11 Yes
11	New development: Require multi-family individual unit metering on new accounts	Water retailers'/ city/county requirement	5	4	2	11 No
12	Existing multi-family individual unit metering incentive	Water retailers'/ city/county requirement	4	4	2	10 No
13	Residential efficient outdoor use education and training programs	Water retailers' sponsor	5	5	3	13 Yes
COMMERCIAL						
14	Top CII users survey program	Water retailers' rebate	5	5	4	14 Yes
15	CII rebates to replace inefficient equipment	Water retailers' rebate	5	5	3	13 Yes
16	Customized CII incentive program	Water retailers' rebate	5	5	3	13 Yes
17	New development require plan review for new CII	Water retailers' requirement	5	5	4	14 Yes
18	CII irrigation water surveys	Water retailers' sponsor	5	5	2	12 Yes
19	Commercial efficient outdoor use education and training programs	Water retailers' sponsor	4	3	3	10 No
20	Turf removal CII/multifamily	Water retailers' rebate	5	5	2	12 Yes

Figure 3-19 Water-saving fixture – waterless urinal
Source: Maddaus Water Management

Figure 3-20 Testing root depth with a soil probe
Source: Maddaus Water Management

Market Penetration

The market penetration for existing customers is the estimated percentage of customers that will be participating in the measure when completed. Estimates are based on measure design and experience from similar measures implemented by other water agencies (see Figure 3-22).

The concept of market penetration can be explained using the example of residential water surveys. If approximately 10,000 residential dwellings exist when a measure begins,

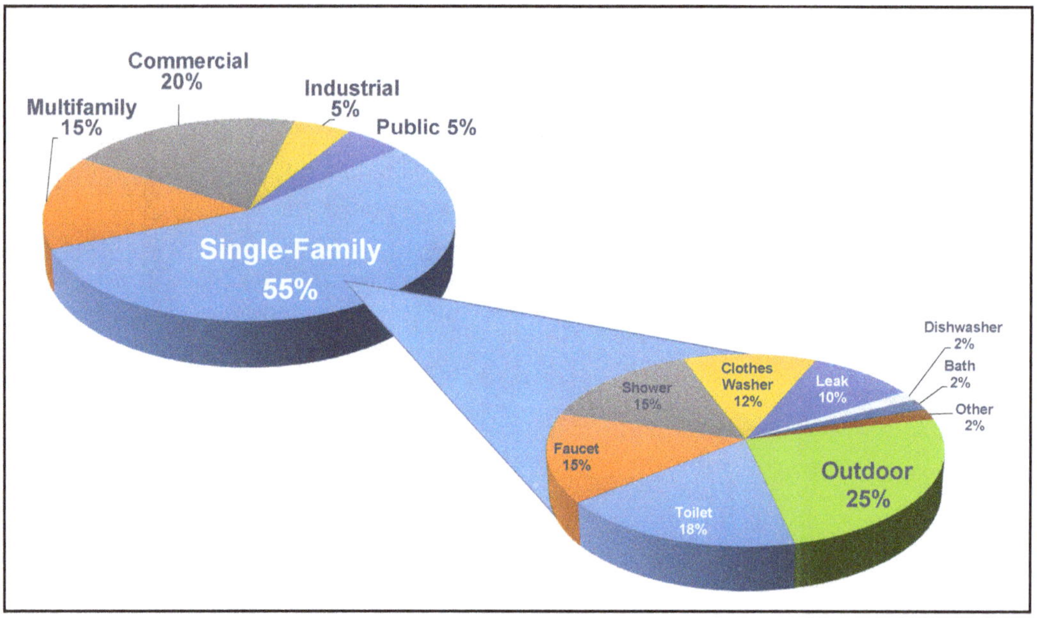

Figure 3-21 Typical overall water sales breakdown by type of customer and indoor water use

Source: Adapted by Maddaus Water Management from WRF (2016)

and the ultimate penetration rate of 10 percent will be reached after three years, 1,000 customers would have participated by the third year. Each year 333 new dwellings would be surveyed until all 1,000 had been audited. Certain measures require maintenance or repetition. For example, surveys would need to be done every year in order to maintain savings because the effects of the surveys may have a limited life. Thus, if water savings from the surveys are assumed to last five years (the life of the measure), then additional surveys (in this case 333) or other appropriate follow-up with prior surveyed homes should be done every year to ensure that water savings are permanent.

Errors in market penetration estimates for each measure can be significant because they are based on previous experience, chosen implementation methods, and projected effort and funds allocated to the measure. The potential error can be corrected, through reevaluation of the measure, as the implementation of the measure progresses. For example, if the market penetration required to achieve the needed savings is more or less than predicted, adjustments to the implementation efforts can be made. Larger rebates or more promotions may be used to increase the market penetration, for example. The process is iterative to reflect actual conditions and helps to ensure that market penetration and needed savings are achieved regardless of future variances between estimates and actual conditions.

In contrast, market penetration for mandatory ordinances can be more predictable. An efficient toilet program could be adopted by ordinance requiring that plumbing is retrofit within a set time or on resale. The program can ensure an almost 100 percent market penetration for affected properties. An example of market penetrations for conservation measures used on an example project are shown in Table 3-10. The market penetration is based on the factors previously described and is inherent in the measure design and cost. Table 3-10 shows illustrative values only and may not be more generally applicable.

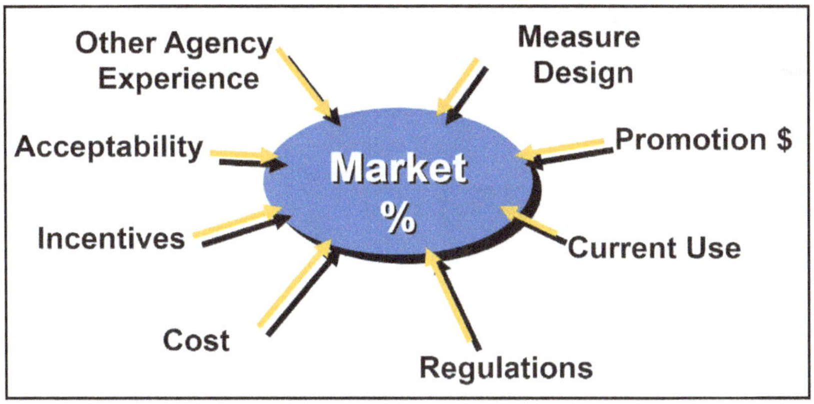

Figure 3-22 Assess market penetration

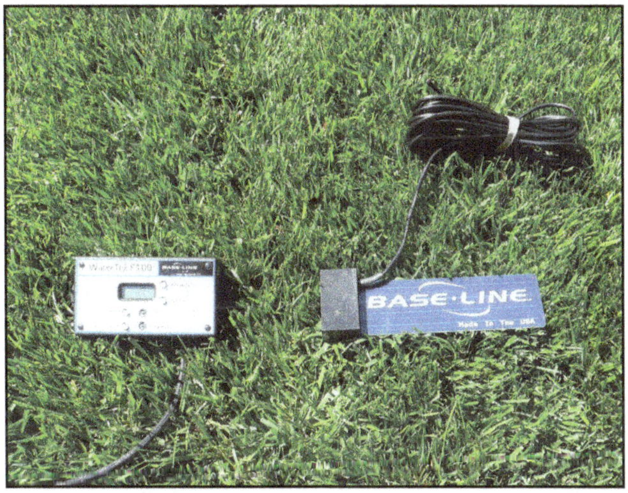

Figure 3-23 Soil moisture sensors have been proven to save water but require adjustment and periodic maintenance

Source: Maddaus Water Management

Example of Unit Water Savings Expressed on an End-Use Basis

An example of end-use water savings, presented in Table 3-10, is expressed as a percent reduction in water use per end use. The percentages only apply to the amount of water identified as the end use, not the entire category of use. Long-term savings are those that are sustainable. Measure life is also shown in the table. When the measure life is exceeded, the water savings erode, unless steps are taken to maintain them, such as replacing an expiring water audit with a new one. Again these percentage water use reductions are illustrative only and should not be considered generally applicable. Figure 3-23 is an example of a device whose useful life may be limited.

Example Comparison of Water Savings by Measure

The examples in this chapter are intended to show the general screening process. When resources allow, utilities should seek out additional tools to assist them. The water savings screening analysis can be performed using an end-use evaluation model, such as the DSS Model (Maddaus 2013), the Alliance for Water Efficiency Tracking Tool BMP Cost-Effectiveness Spreadsheets (AWE 2011), or custom-made spreadsheets.

Table 3-10 Example of End-Use Water Savings and Market Penetration of Conservation Measures Selected for Evaluation (Passing the Screen)

No.	Measure	Applicable Customer Classes	Water Use Reductions for Targeted End Use(s)	Program Length (year)	Market Penetration by End of Program, %	Measure Life (year)
1	Water loss control program: Compute infrastructure leakage index (ILI) on an annual basis	SYS	10%	30	100	Permanent
3	Public information program including school grants	RSF	0.5%	30	100	Permanent
4	Conservation pricing	RSF	15%	30	100	Permanent
5	Residential high-efficiency toilet (HET) rebates	RSF/RMF	63%	5	10	Permanent
6	Residential high-efficiency washer (HECW) rebates	RSF/RMF	34%	5	10	Permanent
7	Single-family outdoor water surveys	RSF	10%	30	15	Permanent
8	Turf removal SF	RSF	25%	30	10	Permanent
10	Mobile home park submetering	RMF	15%	30	90	Permanent
13	Residential efficient outdoor use education and training programs	RSF	10%	20	15	Permanent
14	Top CII users survey program	CII	5%	5	50	Permanent
15	CII rebates to replace inefficient equipment	CII	5% Indoor 10% Outdoor	30	25	Permanent
16	Customized CII incentive program	CII	20%	7	15	Permanent
17	New development: Require plan review for new CII	CII	5% Indoor 10% Outdoor	17	100	Permanent
18	CII irrigation water surveys	CII	15%	16	25	3
20	Turf Removal CII/MF	CII	25%	30	10	Permanent

Notes: RSF = residential single family; RMF = residential multifamily; SYS = system; CII = commercial/industrial/institutional

Table 3-10 shows what a completed first screen of measures might look like, prior to any detailed economic analysis. The determination of the economic feasibility of water conservation programs depends on comparing the costs of the programs to the benefits provided. The example in Table 3-11 is from a city in Southern California. Savings are averaged over a 30-year analysis period and expressed as a percentage of the region's water needs in 2030. This table presents how much water the measures would save if the measures were run on a stand-alone basis – i.e., without interaction or overlap from other measures that might address the same end use(s). Figures 3-24 and 3-25 are examples of commercial and residential washing machines that have different water savings factors.

Figure 3-24 Large commercial washers can be very efficient
Source: Maddaus Water Management

Figure 3-25 The amount of water used by efficient residential washers has dropped two-thirds in 30 years
Source: Maddaus Water Management

Table 3-11 Example Results of a Conservation Measure Savings Evaluation

	Conservation Measure	Example Water Savings in Year 30 (percent of year 30 water needs)
1	Water loss control program: Compute infrastructure leakage index (ILI) on an annual basis	1.32%
3	Public information program including school grants	0.21%
4	Conservation pricing	1.00%
5	Residential high-efficiency toilet (HET) rebates	0.06%
6	Residential high-efficiency washer (HECW) rebates	0.04%
8	Single-family outdoor water surveys	0.06%
9	Turf removal SF	0.32%
10	Mobile home park submetering	0.22%
13	Residential efficient outdoor use education and training programs	0.09%
14	Top CII users survey program	0.01%
15	CII rebates to replace inefficient equipment	0.07%
16	Customized CII incentive program	0.04%
17	New development: Require plan review for new CII	0.03%
18	CII irrigation water surveys	0.01%
20	Turf removal CII/MF	0.30%

Notes: MF = multifamily; CII = commercial, industrial, institutional

REFERENCES

Alegre, H.; W. Hirner; J. Baptista; and R. Parena. 2000. *Performance Indicators for Water Supply Services*. Manual of Best Practice series. London: International Water Association.

Alliance for Water Efficiency (AWE). 2011. Water Conservation Tracking Tool User Guide. http://www.allianceforwaterefficiency.org/Tracking-Tool.aspx, accessed February 13, 2017.

American Water Works Association (AWWA) Water Loss Control Committee. 2003. "Applying Worldwide Best Management Practices in Water Loss Control." *Journal AWWA* 95(8):65–79.

AWWA. 2007. *Water Resources Planning*, 2nd ed. Manual M50. Denver, CO: AWWA.

AWWA. 2016. *Water Audits and Loss Control Programs*, 4th ed. Manual M36. Denver, CO: AWWA.

AWWA Water Loss Control Committee. 2014. Free Water Audit Software, version 4.2. https://www.awwa.org/resources-tools/water-knowledge/water-loss-control.aspx, accessed February 10, 2017.

AWWA Water Loss Control Committee. 2010. "Water Audits: Coming to a Neighborhood Near You?" *Opflow* 36(7):16.

AWWA. 2017. *Principles of Water Rates, Fees, and Charges*, 7th ed. Manual M1. Denver, CO: AWWA.

AWWA Partnership for Safe Water, Distribution System Optimization Program. https://www.awwa.org/resources-tools/water-and-wastewater-utility-management/partnership-for-safe-water/distribution-program.aspx

AWWARF. 2007a. *Evaluating Water Loss and Planning Loss Reduction Strategies*. Denver, CO: AWWA Research Foundation.

AWWARF. 2007b. *Leakage Management Technologies*. Denver, CO: AWWA Research Foundation.

Beecher, J. 2002. *Survey of State Agency Water Loss Reporting Practices*. Final report to AWWA Technical and Educational Council. Indianapolis, IN: Beecher Policy Research.

DeOreo, W.B.; P.W. Mayer; E.M. Opitz; B. Dziegielewski; J.C. Kiefer; W.Y. Davis; and J.O. Nelson. 2016. *Residential End Uses of Water, Version 2 – 4309*. Denver, CO: AWWA Research Foundation.

Georgia Environmental Finance Authority (GEFA). 2017. "Water Loss Technical Assistance." http://gefa.georgia.gov/water-loss-technical-assistance. Accessed February 10, 2017.

Green, D.; and W. Maddaus. 2010. *Water Conservation for Small and Medium-Sized Utilities*. Denver, CO: AWWA.

Lahaye, John. 2016. Personal communication. Marin Municipal Water District, Corte Madera, CA.

Maddaus W., L. Maddaus, M. Maddaus. 2013. Maddaus Water Management International Water Association Manual "Preparing Urban Water Use Efficiency Plans." London, UK: IWA.

Maddaus, M. 2013. "Modeling Innovative Conservation Measures." *Proceedings AWWA Sustainable Water Resources*, Nashville, Tenn.

Thornton, J.; A. Lambert. 2006. "Managing Pressure to Reduce New Breaks." *Water21* Magazine. London: IWA Publishing.

Thornton, J.; G. Kunkel; and R. Sturm. 2008. *Water Loss Control*, 2nd ed. New York: McGraw-Hill.

Vickers, A. 2001. *Handbook of Water Use and Conservation: Homes, Landscapes, Businesses, Industries, Farms*. Amherst, MA: WaterPlow Press.

Water Research Foundation (WRF). 2014. *Effective Organization and Component Analysis of Water Utility Leakage Data*. Denver, CO: WRF.

This page intentionally blank.

Chapter 4

Evaluation of Benefits and Costs

INTRODUCTION

Once a short list of conservation measures has been selected (see chapter 3), further analysis is warranted. Water conservation planners often rely on cost-effectiveness analysis to compare water conservation measures. This type of analysis is a systematic way to evaluate benefits and costs associated with measure implementation. A conservation measure is said to be cost-effective if the present value of the benefits exceeds the present value of the costs. In this manual, the terms *cost-effectiveness analysis* and *benefit–cost analysis* or *benefit– cost analysis* are used interchangeably. This chapter uses benefit–cost analysis. Other less common methodologies may also be used by the reader, such as life-cycle cost analysis, but the end result is the same. Some measures are shown to be more economical than others.

The planner may use benefit–cost analysis to educate the public and decision makers to enable informed decisions. An example case study of a benefit–cost evaluation for a residential survey efficiency measure is provided at the conclusion of this chapter. Performing a benefit–cost analysis sets the stage for the monitoring and future evaluation of the program over time.

The first step in evaluating benefits and cost is to determine the perspective of the accounting to which the benefit or costs accrue. There are three basic perspectives:

- Water utility
- Utility customers (conservation program participants and nonparticipants)
- Community as a whole (social or external)

In other words, the planner must establish the basis for who receives the benefit or must pay the cost and be consistent in the perspectives to have an accurate analysis. Additionally, benefits and costs may be determined separately for the customer and the

Figure 4-1 Replacing front yard turf with low water use landscape (with a utility rebate) is popular and saves water but is expensive for both the utility and the customer

Source: Maddaus Water Management

water utility and then combined for the community as a whole. Figure 4-1 is an example of where costs and benefits of low water use landscaping (it is generally more expensive but saves considerable water) can be attributed to both the utility and the customer. The question becomes: Is the measure cost-effective to both parties, just one, or none?

This chapter will outline the methodology for performing a benefit–cost analysis to compare the value of conservation measures (through demand management) between themselves and compared with water supply augmentation (increased system capacity, new supply projects). Planners may use this analysis to assess the ability of a conservation program to defer, downsize, or potentially eliminate the near-term need for new water supplies. The outcome of this benefit–cost analysis can allow for more accurate planning for future water supply facilities requirements. Quantifying the benefits (cost savings) from a conservation program will also help justify budget requests to fund the program.

OVERVIEW OF BENEFIT–COST ANALYSIS METHODOLOGY

Benefit–cost analysis is a very useful analysis tool, but the results depend on an accurate forecast of the water savings, costs, and benefits. This manual describes the process to estimate costs and benefits using either a streamlined approach (suitable for small utilities) or a detailed methodology.

Many water utilities use benefit–cost analysis to evaluate and select a conservation program best suited to local conditions. This type of analysis requires locale-specific data about water use and demographics. Collect forecasted demographic data from local planning agencies or the US Census (www.census.gov) can be used to forecast water account growth. The detailed methodology for benefit–cost analysis (Figure 4-2) may be described by the following numbered steps (which are further discussed in this chapter):

1. Develop a projection of water use without any additional conservation, including no plumbing code or appliance standards.
2. Prepare a baseline projection of water use including effects of applicable federal, state, and local codes and standards (often called passive conservation).

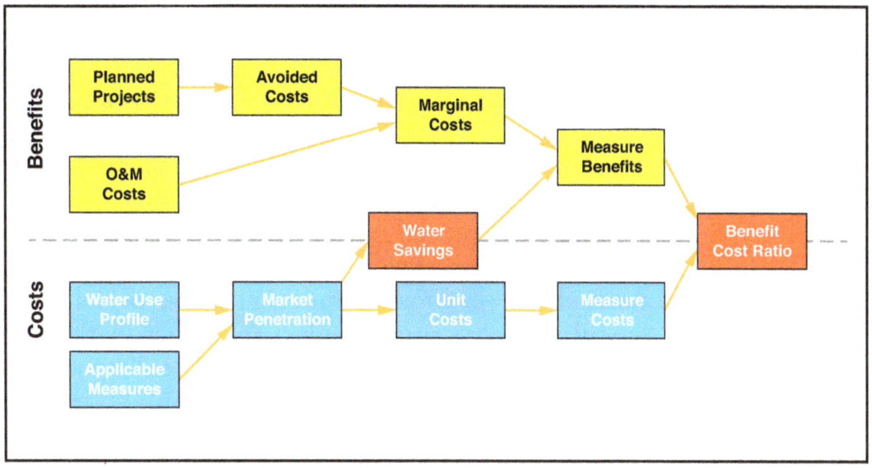

Figure 4-2 Benefit–cost analysis methodology

Source: Maddaus Water Management

3. Based on the profile of use by different customer groups (water use profile), identify all potentially applicable water conservation measures.
4. Screen the measures to select a short list of measures that warrant a benefit–cost analysis (see chapter 3).
5. For each measure, estimate the market penetration (the percent of customers that will participate in the measure).
6. Estimate the average and peak-day water savings by multiplying the number of participating accounts by the measure's unit water savings.
7. Estimate the measure costs by multiplying the number of accounts participating by the measure's unit cost. Convert to the equivalent present value in today's dollars.
8. Identify the benefits to both customers and the utility (water, wastewater, energy). Convert to present value.
9. Identify planned capital projects that might be delayed or downsized by reduced water use and/or wastewater flow.
10. Determine the avoided capital costs associated with the planned water supply (or wastewater capacity expansion) projects.
11. Determine the operation and maintenance (O&M) cost savings associated with reduced water use and wastewater flows.
12. Combine the avoided capital cost and O&M cost savings into the marginal costs, i.e., the cost of the next unit of water.
13. Use the water savings and marginal costs to compute the measure benefits.
14. Compute present value of time stream of benefits and costs.
15. Divide the benefits by the costs and express as a ratio.
16. The measure is cost-effective if the benefit–cost ratio is greater than 1.0.

After the benefit–cost ratio is determined, some utilities also add a percentage bonus for benefits that are difficult to quantify, such as environmental, socioeconomic, and consumer relations. Thus, a measure with a benefit–cost of slightly less than 1.0 might still be undertaken if it is believed that it will result in additional unquantified benefits.

Approach for Small Utilities

For small utilities (fewer than 10,000 connections), a streamlined approach to benefit–cost analysis may be appropriate. While a more detailed evaluation is helpful for informed decision making, especially when a large budget increase is requested, it is not necessary for a successful program. A more informal comparison of benefit and costs may be adequate for the utility to make comparisons and decisions.

The streamlined approach is to compare a planned program measure with another utility's program with similar goals and objectives. This approach can build a budget based on a scaled level of effort such as dollars per person per year cost. For example, take a successful program's annual budget and divide by the corresponding service area population, then multiply the dollars per person cost for the same planned conservation measures by the service area population to determine an estimate for an annual budget request. For example, a neighboring utility has 10,000 connections and a successful school education program for $20,000 per year. The planner may estimate, based on a ratio of 7,500 connections, to budget (7,500/10,000) × $20,000 = $15,000.

This straightforward approach works well when the number and types of utility connections are similar. However, if the neighboring utility has several large industrial customers and many commercial connections, the focus of their program may be on nonresidential conservation measures. It would be shortsighted to then directly compare their efforts to another service area that is predominantly residential. The straightforward approach will likely underestimate costs if the number, general pattern of use, and types of connections (residential, commercial, industrial, institutional) are more than 10 percent different from the comparison utility.

TYPICAL BENEFITS AND COSTS

This section summarizes the major types of benefits and costs from water efficiency to the water utility, whose perspective is the focus of this manual. However, some utilities may also wish to consider customer and society benefit–cost discussed later in the section Other Perspectives on Benefits and Costs. If the measure has a positive (greater than 1.0) benefit–cost ratio for the customer, the customer may be more likely to implement the measure. In addition, benefits accrue from reduced wastewater flows (e.g., lower cost of treatment and delayed facility construction). While these impacts need not be included in the benefit–cost analysis for the water utility, they can be counted for utilities that provide both water and wastewater services. Many water conservation measures also save energy and/or have other benefits. All benefits should be recognized and discussed during the public review phase of efficiency planning. Partnering with wastewater and energy agencies may assist with funding efficiency programs.

Important factors to consider besides a formal benefit–cost analysis are other noneconomic impacts (e.g., water quality improvements caused by less runoff from irrigated landscapes that carry pesticides and fertilizers through stormwater systems, which discharge these contaminants directly to streams and rivers). Quantifying social and environmental benefits so they can be included in the quantitative benefit–cost analysis has long been problematical. They should be considered at least qualitatively.

Reduced water production will allow the utility to reduce costs in the following ways:

- Reduced water purchases from wholesale water agencies
- Reduced O&M costs (energy from pumping [production, treatment, and distribution] and lower chemical use)

- Reduced or deferred treatment plant capital expansion costs
- Reduced water storage costs
- Reduced wastewater processing costs

Water utility cost savings can be significant. Savings depend on the utility's cost of water, which can depend on necessary treatment of the source water. If water availability and quality are limited and new sources are needed, the cost of providing water may rise rapidly. These benefits (cost savings) are based on combined short-term and long-term water savings.

Short-term savings – those related to reduced pumping and treatment – result immediately from efficiency activities. These include the reduced costs of treatment chemicals, energy, and labor and materials required to handle reduced water production.

Long-term savings are those associated with capital facilities and the cost to operate them (i.e., deferred, downsized, or avoided water and wastewater facilities because of reduced demand) or reduced water purchases. These facilities savings include not only the obvious new sources of supply, but also distribution improvements needed for increments of that new supply (e.g., replacement requirements to increase size of distribution system pipelines).

Cost of efficiency programs falls into three broad categories:

1. Implementation costs (paid by the utility) such as staff time, hardware costs, and public and school education materials, and the cost of any monetary incentives that may be offered.
2. The cost to the utility from reduced revenues resulting from decreased demand, normally accommodated by water rate adjustments.
3. Other costs that accrue to customers or other agencies to install and maintain water conservation measures (e.g., Xeriscape™ demonstration gardens, irrigation system improvements).

HOW TO CALCULATE WATER SAVINGS

The following guidance provides details for estimating water savings. An example of estimating water savings from specific measures was provided in chapter 3, specifically in Table 3-10. The formulas give specific details on how to calculate water savings that will result in cost savings (or benefits) from water conservation. These calculations show how to calculate the rate of water savings in millions of gallons per day (mgd) and may also be performed to calculate the volume of savings (e.g., acre-feet or million gallons per year) using life cycles of measures.

To calculate estimated water savings, baseline water use must first be determined for the group of users targeted (e.g., residential customers). Water savings resulting from conservation measures will depend on (1) the reduction in water use as a result of implementing the measure to a target group of customers and (2) the degree of coverage that the measure can achieve (also known as *market penetration*). Generally, this formula is applied to the customer class or classes targeted and is in terms of billed water. The following formula may be used to estimate how much water a specific efficiency measure may save in a given year:

$$E = R \times C \times V \tag{4-1}$$

Where:

E = estimated reduction in water use for the service area (total of customer classes targeted) as a result of the measure for the year of interest, in units such as gallons per day (gpd).

R = fractional water use reduction for customers targeted, as a result of the measure, expressed as a fraction of 1.

C = coverage of the measure for the group of water users under consideration (market penetration) for the year of interest, also called the *installation rate*, expressed as a fraction of 1. For mandatory measures (e.g., plumbing efficiency standards), the C factor is considered 1.0 (100 percent coverage). For voluntary measures, the C factor is much lower. One resource for estimating this value is another utility's experience. Another approach is to set a value for C based on the desired coverage for the program. For example, the utility may decide that a coverage of 20 percent is the goal for the implementation of the measure (e.g., home water surveys), and the efficiency program will be designed to achieve this goal, so in this example the C factor would be 20 percent or 0.20.

V = volume of billed water use in service area (targeted customer classes) without the water efficiency measure, in million gallons per day or year (gpd or mgy).

The fractional water use reduction, which shows the reduction in water use as a result of the measure, can be *estimated* by the formula:

$$R = S / W \qquad (4\text{-}2)$$

Where:

R = fractional water use reduction for the year of interest

S = water savings resulting from the measure, in gpd or mgy

W = average volume of water use without the efficiency measure in place (gpd or mgy), for the year of interest.

In using this equation, it can be applied to the entire service area or one group of users, say single-family residential. So the values used in the equations, such as V, the total volume of water without the measure, needs to apply to the targeted group. In this way the reduction R will also apply to the same targeted group. Likewise, the percent reduction in water use from the measure can be expressed as:

$$P = R \times C \qquad (4\text{-}3)$$

Where:

P = overall percentage of water use reduction from the intervention

R = fraction water reduction for year of interest

C = coverage of measure for group installation rate, expressed as a fraction

For example, assume a town of 10,000 connections intends to do a showerhead replacement program that achieves coverage of 50 percent of the households. The existing showerheads are assumed to flow at 3.0 gallons per minute (gpm). The new showerheads flow at 2 gpm. Showers are assumed to last 8 minutes and showers per capita are estimated to be 0.8 per day. Assume that each connection serves an average of 2.6 people. So estimated showers per connection per day are 2.08.

Shower use volume for existing (3 gpm × 8 min × 2.08)	49.92	gpd/connection
Shower use volume for new (2 gpm × 8 min × 2.08)	33.12	gpd/connection
Water savings from measure	16.8	gpd/connection

The fractional water use reduction is 0.34:

$R = S/W$		0.34	
S	water savings from the measure	16.8	gallons saved per day per connection
W	average existing water use for showering per connection per day	49.92	gallons used per day per connection
V	volume of water without the water efficiency measure = (49.92 × 10000 × 365)/1,000,000	182.21	mgy

The reduction in water use from installation of water efficient showerheads is:

$E = R \times C \times V$	30.66	mgy

The overall percentage of water use reduction from the intervention would be:

$P = (R \times C) \times 100$		17	percent
R	fraction water reduction for year of interest	0.34	
C	percent coverage of measure for customer group or installation rate	0.5	

To compute the benefits of water conservation using these formulas, the planner will need to estimate the water savings that may result from each of the conservation measures. Figure 4-3 is an example of an efficient spray rinse nozzle that published sources indicate can save considerable water in a commercial kitchen. About 970,000 food service establishments in the United States use approximately 51 billion gallons of water each year to rinse dishes with pre-rinse spray valves. In fact, pre-rinse spray valves can account for nearly one third of the water used in the typical commercial kitchen. The US federal standard for commercial pre-rinse spray valves is 1.6 gpm, but manufacturers have now developed models that use significantly less water. Replacing one pre-rinse spray valve with a WaterSense®–labeled model that uses 20 percent less than the US federal standard (USEPA 2017) can save a typical commercial kitchen more than 7,000 gallons of water per year.

There are numerous resources to assist with estimating water savings. See references at the ends of chapters 2 and 3. Sources include:

Figure 4-3 Water savings from replacing inefficient kitchen spray rinse valves – well documented in the literature and a relatively low cost measure

Source: Maddaus Water Management

- *Residential End Uses of Water Version 2* (WRF 2016)
- *Commercial and Institutional End Uses of Water* (Awwarf 2001)
- *BMP Costs & Savings Study: A Guide to the Data and Methods for Cost Effectiveness Analysis of Urban Water Conservation Best Management Practices* (CUWCC 2005)
- *Handbook of Water Use and Conservation: Homes, Landscapes, Businesses, Industries, Farms* (Vickers 2001)
- *Water Conservation for Small and Medium-Sized Utilities* (Green and Maddaus 2010)
- *Preparing Urban Water Efficiency Plans, A Best Practice Guide* (Maddaus et al. 2013)
- California Urban Water Conservation Council's Memorandum of Understanding that contains a list of Best Management Practices http://www.cuwcc.org/Resources/BMP-Resources
- Arizona Department of Water Resources list of Reasonable Conservation Measures
- The US Environmental Protection Agency's WaterSense program at https://www.epa.gov/watersense
- Water Saver Home Web Site, http://www.home-water-works.org/calculator
- Full program savings assessments and pilot test results published in Journal AWWA and in Annual and AWWA Water Sources Conference Proceedings
- Data from other nearby water utilities or the state agency tasked with overseeing urban water demand (e.g., California Department of Water Resources).

HOW TO DETERMINE THE BENEFITS OF CONSERVATION MEASURES

Savings to the utility result from avoided costs (the benefits from implementing conservation measures that achieve water savings). This section describes the three principal ways that avoided costs accrue: (1) reduced water purchases (if the utility is a wholesale customer of another water purveyor); (2) lowered O&M expenses; and (3) delayed, downsized, or eliminated capital facilities.

If a utility buys water from a wholesaler and has no sources itself, the supply-side benefits to the utility from saving water are easy to compute: It is the reduced purchase cost. If the utility needs less, it buys less, and the benefits can be readily determined from purchase agreements. For the wholesaler that provides the water or for utilities that supply their own water, deferring or downsizing projects is a major source of potential benefits for the conservation program. If water use is growing and will exceed the capacity of supply sources and water treatment and distribution facilities, capital expansion projects will be needed. Unfortunately, the benefit–cost analysis is often hindered by the lack of long-term water supply capital facility plans (giving types of projects and cost estimates). Sometimes capital facility plans are only determined three years in advance to correspond to water rate reviews. Large water supply projects normally require a long lead time to go through environmental review, obtain permits and funding, arrange for engineering, and so on.

If capital facilities appear to be needed and the plans do not cover the normal conservation planning period (usually 20 years), the costs of these unplanned facilities should be estimated. Supply projects are normally designed to provide for growth over a 20- to 50-year period. Similarly, expansion of water treatment projects is designed for growth over approximately 10 years. There are general cost guidelines available from various sources for estimating project costs sufficiently and accurately for conservation planning. In this way, the costs to accommodate planned growth for 20–30 years can be estimated, and the conservation analysis can proceed.

Cost Savings from the Reduced Purchase of Water

A straightforward calculation results in the average annual unit cost of purchased water from a wholesaler using the following expression:

$$AWC = UPW \times UCPW \tag{4-4}$$

Where:

AWC = annual water cost
UPW = units purchased annually
$UCPW$ = unit cost of purchased water

The planner can calculate the amount of cost savings by multiplying the unit cost of purchased water by the units of water savings estimated from conservation measures. An added level of detail can be used if a higher cost is charged in peak-use period (e.g., high irrigation season). Then, the average cost during this period (typically a few months' time) is divided by the units of purchased water over the same time span. This unit cost of peak-period purchased water is multiplied by the water savings from conservation measures targeting water reductions during that period (commonly outdoor irrigation conservation measures).

Cost Savings from Reduced O&M Expenses

Because lowering demand results in less water produced, conservation measures can reduce expenses dependent on amount of water produced or variable costs for utility operations, such as energy and chemical costs. In addition, some fixed costs may be associated with these variable costs of energy and chemical usage and may be included, if warranted. Only the variable costs that are attributed to water efficiency activities are included in the calculation of avoided costs shown here.

To estimate the variable cost of energy ($/unit of water), use the following formula:

$$VUCE = [AEC - (12 \times MFEC) - (ECNP)] / UWU \qquad (4\text{-}5)$$

Where:

VUCE = variable unit cost of energy

AEC = annual energy bill (cost)

MFEC = monthly fixed charges for energy

ECNP = energy costs not related to water production, that is, those costs independent of actual water production such as building heating, cooling, lighting, and process equipment use. These costs should be included unless the water production is lowered to the extent that facilities (e.g., certain buildings or pieces of equipment) are not used, which would rarely be the case.

UWU = annual units of water used

Cost savings are calculated by multiplying the variable unit cost of energy by the units of water saved per year as a result of an efficiency measure.

In most cases, costs associated with chemicals are variable because the chemicals are added based on flow with very few fixed costs. To calculate the variable cost of chemicals ($/unit of water), use the following formula:

$$VUCC = [ACC - CCNP] / UWU \qquad (4\text{-}6)$$

Where:

VUCC = variable unit cost of chemicals

ACC = annual chemical bill (cost)

CCNP = chemical costs not related to production (e.g., delivery charges unless reduced)

UWU = annual units of water used

Cost savings are calculated by multiplying the unit cost of chemicals by the units of water saved per year as a result of a conservation measure. The benefits derived from wastewater operations for energy and chemical savings can be calculated in a similar manner.

Cost Savings from Delayed, Downsized, or Eliminated Capital Facilities

Water efficiency can affect both the requirements for current operations, expansion of existing facilities, and planned new facilities. Typical design horizons are 10 to 20 years. To evaluate the impacts of efficiency, both peak-day and average-day water use must be considered. Peak-day water use usually occurs on or near the warmest day of the year for the community when outdoor irrigation has the highest demand of the season. Most capital facilities are designed to meet peak demands in some future year. Although indoor conservation measures will reduce average-day and peak-day demands, savings in landscape, cooling water, and other summer uses will have greater effects on reducing the peak. In cities with hot or arid climates, peak-day to average-day ratios of 2.0 to 3.0 are common. In humid or colder climates, peak-day ratios of 1.2 to 1.7 are common. The peak-day ratio can be determined by comparing utility water production records using the following formula:

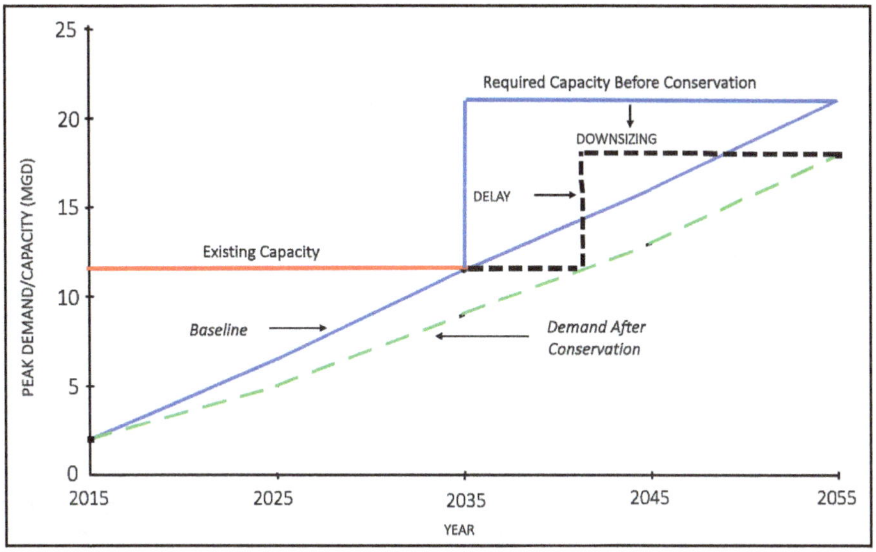

Figure 4-4 Example of delaying and/or downsizing a capital facility
Source: Maddaus Water Management

$$\text{Peak-day ratio} = \text{Maximum day production} / \text{average day production} \quad (4\text{-}7)$$

The timing of capital facilities depends on the rate of growth in peak demand and the capacity of existing facilities. If the planned facilities are dependent on the growth of water demand, reduction in future water use can affect the timing of construction of these facilities. Figure 4-4 illustrates how water conservation could affect the timing of capital facilities. In this case, a water treatment facility needed in 2025 could be delayed about seven years. In the example shown, demand reduction would reduce peak-day demands by about 20 percent. The resultant dollar savings to the utility are the difference in the present value of the costs associated with building the facility in 2032 instead of 2025.

A utility's efficiency program would reduce peak-day water use by 15 percent. Cost savings to the utility are the difference between building the plant at two different points in time (less debt service), plus the elimination of operating expenses for the years of delayed construction.

If demand is leveling off as growth slows down, reducing demand may reduce the need for the last expansion. In this case, the last expansion can be downsized or delayed. The capital cost savings associated with a smaller facility can be converted to present worth and added to other conservation benefits.

Information on the timing and sizing of capital facilities can often be found in the utility's capital facility plan, water supply plan, and/or water master plan. Unfortunately, sometimes the capital facilities are only identified a few years in advance, and projections of needed facilities must be made using demand projections and the design criteria.

Water System Design Criteria. New water facilities may present an opportunity to downsize or postpone expansions if impacts of water conservation are considered at the same time. This can occur if the design of the facility is dependent on water flows. Table 4-1 shows typical design criteria for water facilities that may be affected by reduced consumption. Reduction in average-day demand affects how much water must be developed, or imported and stored, prior to treatment and use. Reduction in peak-day demand affects the sizing and timing of water treatment plant expansions and treated water storage. Water pipelines and pumping stations are affected by peak-hour pumping. Peak hour

Table 4-1 Water System Elements Affected by Conservation

	Design Criteria Based on			
System Element	Average Day	Peak Day	Peak Hour	Fire Flow
Source water acquisition	✓			
Raw water storage	✓			
Water pipelines		✓	✓	✓
Water treatment plants		✓		
Pumping stations			✓	✓
Treated water storage		✓		✓

Source: Maddaus and Maddaus 2014

Table 4-2 Wastewater System Elements Affected by Conservation

	Design Criteria Based on		
System Element	Average Dry Weather Flow	Peak Wet Weather Flow	Solids Loading
Collection systems		✓	
Interceptors		✓	
Treatment plants		✓	✓
Disposal to receiving water		✓	
Land disposal	✓		

Source: Maddaus and Maddaus 2014

is dependent on customer peak-hour demands plus required fire flows. The latter is based on the type of land use to be protected.

Wastewater System Design Criteria. Table 4-2 shows the impacts of conservation (wastewater flow reduction) on design of new facilities. Design criteria for land disposal systems are volume dependent. Most facilities are based on peak wet weather flow, which can benefit from infiltration/inflow (I/I) control programs but are little affected by conservation programs, which save much less water than I/I contributes.

Other Benefits

Other benefits that sometimes are significant and possibly can be quantified:

- Lower withdrawals from supply sources (more water remains in rivers and aquifers)
- Lower discharges of treated wastewater to receiving waters
- Lessened construction environmental impacts
- Creation of water conservation jobs
- Customer savings in energy bills
- Reduced greenhouse gas emissions

In-depth descriptions of the methodology for calculating the economic benefits that the utility will realize from the delay, downsizing, or elimination of capital facilities is available in *Impacts of Demand Reductions on Water Utilities* (Awwarf 1996).

How to Calculate Avoided Costs from Downsizing, Delaying, or Eliminating a Water Supply Capital Project

The basic methodology in this section expands on the concept illustrated in Figure 4-4 and computes the economic benefit using the following simplified formulas:

If the project is downsized:

$$\text{Cost savings} = (\text{cost at original size}) - (\text{cost at reduced size}) \tag{4-8}$$

Cost in Equation 4-8 includes both capital and life-cycle O&M costs.

If the project is delayed:

$$\text{Cost savings} = \frac{(\text{cost in original year})}{(i+1)^m} - \frac{(\text{cost in delayed year})}{(i+1)^n} \tag{4-9}$$

Where:

m = number of years until the planned project is built

n = number of years until the delayed project is built

i = discount rate used in present value calculation

Note: $n - m$ is the delay in years.

Cost in Equation 4-9 includes both capital and life-cycle O&M costs.

If the project is eliminated:

$$\text{Cost savings} = \frac{\text{capital cost of planned project}}{\text{(in present value [today's] dollars)}} \tag{4-10}$$

DETERMINING THE COSTS OF CONSERVATION MEASURES

This section describes two principal types of costs to the utility for undertaking efficiency programs: (1) direct utility costs for implementation, such as in-house staff costs and any contracted costs (where a private contractor performs some of the work); and (2) how to deal with decreased water revenues.

Direct utility costs are the sum of in-house staff costs and contracted costs (if work is contracted out) and can be calculated as:

$$\text{Direct utility costs} = \text{in-house costs} + \text{contract costs} \tag{4-11}$$

Where:

In-house costs = administrative costs + (field labor hours × hourly rate [including overhead]) + (unit costs × number of conservation measures or devices) + publicity costs + evaluation (or follow-up costs)

Contract costs = administrative costs + (number of events [or sites]) × unit cost per event (including program unit costs) + publicity costs + evaluation (or follow-up costs)

Administrative costs include the staff time required to oversee field staff, contractors, consultants, or contracted field labor. Administrative costs will be higher when launching a new program or with large consultant contracts. Administrative costs are typically 10 to 30 percent of implementation costs for the total measure.

Field labor costs include staff time to conduct efficiency program work in the field, such as water audits/surveys, leak repair, fixture installation, follow-up site visits, and door-to-door canvassing.

Unit costs are based on cost per device or measure basis or as a cost per participant. Examples include retrofit kits, water audit programs, and rebate programs. Small programs typically have higher unit costs than larger programs because of bulk purchase discounts and fewer participants.

Publicity costs are the costs of public outreach to educate customers through local media, including radio and television spots, local newspaper advertisements, flyers, bill inserts, billboard and bus advertising, theater slides, customer workshops and seminars, and special demonstrations (booths at community events). Larger utilities often employ public relations professionals to handle this aspect of their efficiency program for maximum effect, but this is not necessary for smaller programs. Costs will be roughly proportional to the number of customers contacted.

Evaluation and follow-up costs include record keeping to quantify savings and monitoring through follow-up contact with participants to assess how well measures are performing and whether program goals are being achieved. Costs from these follow-up activities may include staff time to conduct surveys that assess customer participation and satisfaction, including changes from a baseline survey on attitudes. Market penetration studies (more common among larger utilities) may be conducted to assess future means for better targeting implementation of the measure.

The best sources of information are other utilities that have conducted similar programs. Costs can be expressed on a unit basis (for example, $/dwelling unit or $/survey/audit) and then transferred to another utility's service area, accounting for economies of scale (e.g., any bulk purchase discount or larger number of participants that would drive costs down) for different size programs.

Figures 4-5 and 4-6 illustrate examples of measures whose costs and benefits may not be published but can be estimated.

Costs of Decreased Water Revenues

There is a direct correlation between lower water use and less revenue. The cost of decreased water revenue is the cost per unit of water multiplied by the units of water conserved. Generally, this revenue reduction is small and occurs over a long period of time, allowing the utility to incorporate these changes into budget forecasts and to redesign rate structures. Typically, cost-effective (benefit–cost ratio above 1.0) efficiency programs save 0.5 percent to 2 percent of annual water use, and by the same accord, reductions in water revenues per year are the same over the life of the program. This amount has historically been less than inflation in other utility costs. The short-term savings from conservation measures that reduce production costs (energy, chemical, and treatment costs) help to offset revenue decreases. Periodic rate adjustments can recover the inflation in utility costs in addition to recover any revenue reduction; thus the actual economic impact is insignificant. The primary concern of utility decision makers over reduced revenue can frequently be avoided by incorporating estimated conservation program savings into future demand forecasts and rates prior to program implementation.

Reduced revenue is a primary concern of utility decision makers and should be assessed carefully and explained fully. Decreased water revenues can mean less funding for new capital facilities O&M. However, these costs are often offset by a reduced need to build, operate, and maintain future facilities.

Take a situation in which no conservation is implemented by argument – no conservation is implemented by the customer, and the utility must build another increment of supply. Now, rates need to go up to $3.00 per 1,000 gallons to be able to continue to supply that customer with the same 10,000 gallons. The customer's bill is $30 a month, and they have not done any conservation. The services they receive from the utility are exactly the same. Now, say that the customer decided to implement conservation to lower their bill and they reduce water use to 8,000 gallons per month. Their monthly bill is now $24. The utility's costs may not go down because this customer saved 2,000 gallons of water – the capital investment for the incremental supply must be paid for. If everyone else also implements conservation, the utility would need to raise rates even higher to generate sufficient revenues to pay for the new supply.

The bottom line is that water rates and revenues are complicated and must be studied when a conservation program is implemented to ensure equity and revenue stability. Chapter 7 has additional discussion on how utilities can balance revenue concerns and rate setting with conservation goals.

HOW TO PERFORM A BENEFIT–COST ANALYSIS

Up to this point, the planner has gathered information on how to calculate benefits and costs. The goal now is to combine this information into a formal benefit–cost analysis from the perspective of the water utility. For guidance, Tables 4-3 through 4-5 provide actual examples of how to perform a benefit–cost analysis for a residential water survey efficiency measure.

Benefit–cost analysis will inform the planner, decision makers, and the public whether the proposed measures are economically efficient, or if the benefits are greater than the costs. The larger the benefits, which include water savings, and the smaller the costs of the measures, the more economically attractive the measures are to the water utility. Later in this chapter, immediately following determination of benefit-to-cost ratio from a utility perspective, is a more detailed discussion of other perspectives and considerations.

Benefit–cost analysis requires careful attention to detail and is a central responsibility of planners at medium-size and large utilities. Planners perform benefit–cost analysis to justify significant budgets or as part of an effective water supply planning process. Smaller utilities may elect to calculate the cost of water saved, as described in a following section, and select measures based on only costs.

A positive benefit–cost ratio will not always be the final deciding factor. Some measures are implemented independent of an economic evaluation. A good example is public education programs, which are often thought of as the glue that holds the efficiency program together. When performing a financial assessment, public education is difficult to quantify in terms of direct water savings and, as a result, rarely has a positive benefit–cost ratio. However, public information and education programs are critical to assist with the success of all measures by building the conservation ethic in customers. In general, most utilities will ramp up their efficiency program over time and package education costs with other cost-effective measures so that the total combined conservation program has a positive benefit-to-cost ratio.

How to Determine the Benefit-to-Cost Ratio

This is a standard means of analyzing different alternatives, and numerous economics textbooks present several methods for estimating the costs and benefits of a potential alternative, in this case an efficiency measure.

As an overview, the method calculates the ratio of the present value (today's dollars) of benefits to the present value of costs. Present value of a future cost or benefit (payment) is the amount of money needed today to make that payment in a future year, given that today's money will earn interest between today and when the payment must be made. It is a similar concept to buying a US savings bond today at a discounted amount that will mature to the face value in some specified future year.

If the ratio is greater than 1.0, the benefits outweigh the costs and the measure is considered feasible (or economically efficient). The following formula shows the basis for benefit-to-cost ratio:

$$\text{Benefit–cost ratio} = \frac{\dfrac{\text{sum of benefits (\$) in year } (t)}{(1+i)^t}}{\dfrac{\text{sum of costs (\$) in year } (t)}{(1+i)^t}} \tag{4-12}$$

Where

i = selected discount rate, as a decimal (5% = 0.05)

t = year in evaluation period

Estimating the Cost of Water Saved

The cost of water saved is a useful number that is relatively easy to calculate. It is commonly expressed as dollars per million gallons, cents per 1,000 gallons, or dollars per acre-foot and can be used as a simple comparison to see if conservation measures are less expensive than new sources of supply. Although the calculation is somewhat simple, the reader is cautioned that the lifetime of each conservation measure and the lifetime of a water supply project are usually different, and thus these should be normalized prior to comparison. This can often be done by renewing the shorter lifetime measure until it approximately equals the lifetime of the longer one. There is no standardized formula for calculating the unit cost of water saved, but the following is suggested:

$$\text{Unit cost of water saved (\$/unit of water)} = TC/V \tag{4-13}$$

Where

TC = present value of the total efficiency program costs over planning period (dollars)

V = total volume of water saved (units) over the planning period (e.g., per acre-foot or per 1,000 gallons)

OTHER CONSIDERATIONS

Looking only at the economic benefits and costs to the water utility may lead the planner down the path of selecting measures only with very high benefit–cost ratios. However, it is vital to keep the big picture in mind and make sure that the overall water savings goal of the program is met. The goal could be annual water savings or peak-day water savings, which would lead to different portfolios of measures. High benefit–cost ratios may only lead to small total water savings and result in using available dollars that would be better expended on implementation of measures that achieve more total water savings or save on peak-day water savings, depending on program goals. Often, packaging measures that target certain customer sectors provide economies of scale in marketing and administration, making the program more cost-effective than the individual components would appear. Combining measures with high benefit-to-cost ratios with some that produce more savings but have lower benefit-to-cost may in some cases be an attractive option.

The City of Austin, Texas, case study in appendix A gives some other perspectives on program design going forward.

OTHER PERSPECTIVES ON BENEFITS AND COSTS

It is crucial to review the conservation measures from other perspectives, namely the customer and other organizations (e.g., wastewater, stormwater, or energy utilities). For example, for conservation measures such as efficient showerheads that result in hot water savings, customers receive significant cost savings on their energy bill in addition to savings on their water bill (if metered). In this case, secondary beneficiaries result from cost savings to the wastewater and energy utilities.

Another consideration is to eliminate free-riders from the analysis. A free-rider is a customer who takes advantage of an incentive even if they would have done the measure in the absence of one. By taking the incentive, the customer adds to the cost of the program but does not increase the benefits. This can be incorporated by reducing the estimated percentage of free-riders. For example, if the customer takes a rebate to replace a fixture but that is the only type of fixture they can buy because of plumbing codes, then the free-ridership could be quite high, possibly over 50 percent. On the other hand, if the fixture qualifying for the rebate is one of a number of choices but cost-wise becomes attractive because of the rebate, the free-ridership could be quite low. The general guidance is to stop incentive programs once the desired outcome has been reached, as required by law.

Where potential conservation measures benefit more than one utility or city/county department, the cost savings to each may be quantified and used as means for negotiating a partnership to co-fund efficiency programs. At a minimum, combined messages are especially effective when trying to stretch dollars spent on public information campaigns. Some examples of potential partnerships include the following:

Stormwater Utilities

Reductions in outdoor irrigation prevents runoff containing pesticides and fertilizers and are beneficial to stormwater utilities governed by discharge water quality limits (sometimes difficult to meet in dry seasons). Irrigation peak-day water use is also the most significant capital cost to water utilities, given that system capacity expansions are commonly triggered when demand reaches 90 percent of peak system capacity.

Solid Waste Utilities

Less irrigation and use of appropriate (more native) and dense landscapes result in less green waste, which saves on trucking and disposal costs. However, in some cases, solid waste utilities have bioreactors that use green waste to generate energy, and reduced waste would lower the benefit to these utilities.

Wastewater Utilities

Reduction in indoor water use leads to the generation of less wastewater, which results in cost savings to wastewater utilities (e.g., lower energy and chemical usage on the average dry-weather flow that is mostly indoor water use). Given that new wastewater treatment facilities are hydraulically sized to meet wet-season conditions and mechanically sized to treat the workload, water efficiency usually does little to assist in deferring wastewater capital projects. There are exceptions to the general rule where the volume discharged matters (disposal limitations). Also, many wastewater utilities are looking to recycle treated wastewater by irrigating landscapes with reclaimed versus potable water. This allows water utilities to reduce peak-day irrigation demands, conserve system capacity, and potentially serve additional customers.

Energy Utilities

These benefits may be twofold. Water utilities that produce less treated water conserve energy associated with treatment and pumping. In addition, customers who install more efficient fixtures and appliances that use hot water (showerheads, faucet aerators, clothes washers, dishwashers) also reduce energy usage. When hot water use reductions correspond with peak use periods, energy utilities may be able to defer building new capacity.

Customer Benefits

The following discussion focuses on how to estimate savings to the customer that will result from lower utility bills for water, wastewater, and energy.

Water Utility Bill Savings. Assuming the customer is metered with a volumetric billing rate, the estimated customer water savings from implementing different conservation measures can be multiplied by the customer's water billing rate. This calculation can become complicated, depending on the type of water rate system in place. In cases where the customer pays a regular fee for water service regardless of usage, there is no economic incentive to conserve water.

When utilities charge different rates to different customer classes, the water savings need to be calculated for each customer class (e.g., residential, commercial, industrial, institutional). In the case of different rates within customer cases, the simplest procedure may be to find and calculate the "average" customer in each class and calculate the "average" customer in each class. To determine the average billing rate, the total water consumption for the customer class should be divided by the number of customers in the class. When matched to the appropriate rate, multiplying by the corresponding water savings estimate gives the cost savings for the average customer. This can be done for the peak month usage (commonly July and August) and the minimum month usage (commonly January and February) to estimate a range of average cost savings for an average customer. This information can be useful in public education efforts, for example, as content for bill inserts. However, the combined cost savings with other utility bills will make a stronger case for efficiency.

For the customers throughout the service area, the average water savings per customer class multiplied by (1) the total number connections for that type of customer and (2) the unit cost per 1,000 gallons will provide the total estimated cost savings to average customers for the efficiency program as a whole.

Wastewater Utility Bill Savings. Customers that are billed based on a flat rate will not realize any benefits from conservation measures that generate less wastewater. Residential customers will benefit from conservation if billed according to a fixed monthly charge plus a unit charge per unit of water use. If this billing method is used, wastewater bill savings can be estimated by multiplying indoor residential water savings estimates by the unit charge. This method is appropriate because reduction in wastewater flows is approximately equal to reductions in indoor water use. Nonresidential customers are typically billed with a fixed charge and unit charge based on water usage, so the same method for estimating cost savings is appropriate.

Energy Utility Bill Savings. Reductions in hot water use translate into direct energy savings from less water heating. These cost savings are significant for both residential and nonresidential users. Most energy utilities charge a basic rate followed by *block rates* for higher energy usage. The marginal energy cost for different customer classes can be found by contacting the local energy provider and asking for the rate of the highest block for the average customer or, alternatively, the marginal cost for energy for that customer class. The average customer probably uses energy at all block levels and so energy saved will be at the highest rate, but this assumption should be verified with the local energy provider. Additional information on the type of energy (electricity, natural gas, or propane) most commonly used will lead to the fuel efficiency of the water heating unit and the resultant water heating capacity in each gallon of water. An assumption for an average length of shower will provide a volume of heated water saved. Similarly, for clothes washing, an average number of loads washed and savings from a water efficient machine will lead to energy savings. The local energy provider can assist with this evaluation. The energy units saved can be multiplied by the cost per unit energy (highest rate) to determine the total cost savings to customers.

Solid Waste and Stormwater Savings. Because customers typically pay a flat fee for solid waste disposal service and do not pay directly for stormwater service, neither may provide savings on customer utility bills. In areas where customers pay by volume of green waste, some reductions in volume can produce savings. Nonutility savings from less investment in these types of services is not considered economically significant to include in this analysis of customer savings.

Tax Credits. Certain states, including Texas, offer sales tax and property tax exemptions for installation of certain water conservation equipment such rainwater harvesting and water reuse.

Environmental and Other Benefits

Environmental benefits can take many forms and be direct or indirect. Examples of direct benefits include the lessened impact on endangered species or biological diversity from reduced water withdrawals from a surface water supply. Examples of indirect or less obvious effects include reduced greenhouse gas emission from power plants because of less pumping energy to extract groundwater. Other benefits can include secondary employment for recreation, aesthetics, and conservation services that benefit the local economy. Some of these environmental and social benefits are now beginning to be better quantified.

Customer Costs

Customers that want to become more water efficient will typically incur costs. These costs commonly include an incremental increase in cost to purchase more efficient appliances (e.g., clothes washers). For example, a customer may replace a clothes washer at the end of its useful life. A more efficient machine may cost $200 more than a less efficient machine. The cost to the customer in this example is not the full cost of the washer, but the incremental cost of a more efficient model. Some utilities offer incentives to customers who purchase more efficient fixtures or appliances to reduce or eliminate the cost difference between less and more water efficient models. In addition, incentives may assist with the customer making a purchase of a new more efficient model in lieu of continuing to repair an old inefficient model. Incentives sometimes help offset installation costs as well (e.g., a plumber's installation of a water efficient toilet). Obviously, if costs are too high, customers may not be willing to participate.

Increased costs may also occur for commercial or industrial facilities where installation of water saving equipment increases O&M expenditures. These customers normally use a simple payback analysis on the time to receive a return on an investment, typically two to three years.

The Tampa, Florida, case study in appendix A gives an example of the results of evaluating water savings, the cost of water saved from conservation measures, and benefits based on avoided costs.

EXAMPLE: BENEFIT–COST CASE STUDY FOR RESIDENTIAL SURVEY MEASURE

In this example, a trained water utility surveyor (or auditor) visits single-family homes to conduct tests and recommend ways to save water (the water-saving measure). The homeowners take voluntary actions to conserve water based on these recommendations. Figures 4-5 through 4-6 illustrate components of a residential water survey.

This case study is from a water utility's perspective and is simplified to evaluate only one efficiency measure and a one-year investment in that measure. The benefits (water savings) over the life of the measure are evaluated against the one-year cost of implementing this measure. If the benefit–cost ratio is greater than 1.0 (that is, the benefits are greater than the costs), the measure can be considered cost-effective and a good investment for the utility.

The water utility and the impact of the water saving measure in this example have these characteristics:

Total daily water produced (gal) = 18,500,000

Average daily total single-family residential water use (gal) = 8,400,000

Number of single-family homes = 28,000

Average water use per home (gpd) = 300

Number of homes per year implementing the water saving measure = 1,000

Program length is 10 years.

Each participating home saves 10 percent due to the water saving measure.

Each participating home saves 30 gpd due to the water saving measure.

Each participating home will implement the measure for five years.

Life of the measure is five years.

Planning period for data analysis is 25 years.

Step 1 – Lifetime Water Savings

The first step is to calculate the lifetime water savings from implementing this water saving measure. Use Eq. 4-1 to calculate the estimated annual water savings. In this case, calculate the savings in gallons per day rather than million gallons per year by using this value in the formula.

$E = R \times C \times V$

R (reduction in water use) = 0.1

C (percent coverage) = 1,000/28,000 = 0.0357

V (volume of use before savings) = 8,400,000 average daily water usage in gallons

Figure 4-5 Testing showerheads – a normal part of a residential water audit
Source: Maddaus Water Management

Figure 4-6 Optimizing irrigation can involve a sprinkler catch-can test
Source: Maddaus Water Management

$$E \text{ (estimated reduction in volume of use)} = 0.1 \times 0.0357 \times 8{,}400{,}000 = 30{,}000 \text{ gpd}$$

Calculate the total savings at the end of the measure life (five years) = 150,000 gpd (30,000 gpd for each of five years).

The total savings in the last year of the program (tenth year, see Figure 4-7) = 150,000 gpd (net savings remain constant after five years due to decay).

The total annual savings decay after the end of the program (after year 10, Figure 4-7) = 150,000/5 = 30,000 gpd.

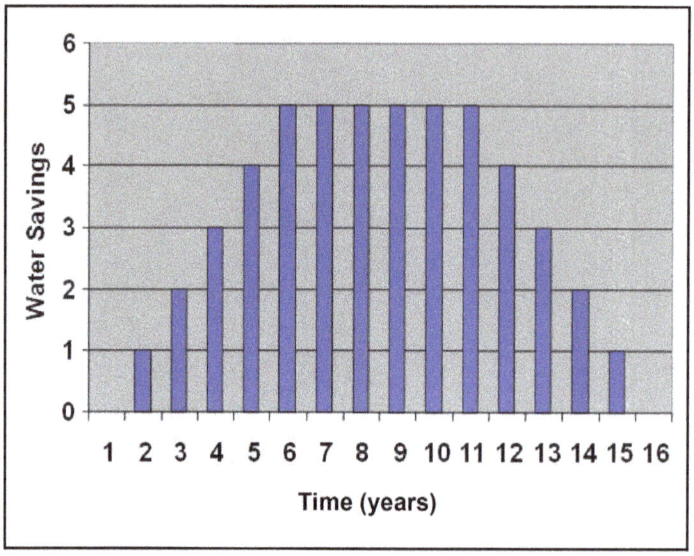

Figure 4-7 Estimated water savings for a 10-year versus ongoing residential water survey program

Note: The savings decay at the end of the five-year measure life, unless the program continues.

The lifetime savings is calculated based on a 25-year planning period that was chosen for the example analysis.

Lifetime savings,
million gallons (MG) = (E × 365/1,000,000) / (1/measure life)
= (E, gpd × 365 [days per year]) / 1,000,000 gal/MG/(⅕ year)
= 54.75 MG

Step 2 – Costs to Develop and Implement the Water Saving Measure (Table 4-3)

Calculate the direct utility costs using Equation 4-14 as follows:

$$\text{Direct utility costs} = \text{in-house costs} + \text{contract costs} \qquad (4\text{-}14)$$

Where:

In-house costs = administrative costs + (field labor hours × hourly rate [including overhead]) + (unit costs × number of conservation measures or devices) + publicity costs + evaluation (or follow-up costs)

Contract costs = administrative costs + (number of events [or sites] × unit cost per event [including program unit costs]) + publicity costs + evaluation (or follow-up costs)

Step 3 – Benefits of Water-Saving Measure (Table 4-4)

The value of the benefits of the water saving measure are the sum of the avoided costs. Three main cost categories comprise this total: reduced water purchases; lower operation

Table 4-3 Summary of Direct Cost Calculation Example

			Value (Example)
1.	Administration costs	a. Staff hours	400
		b. Hourly cost, $/hour	50
		c. Annual cost, $/year, Line 1a × 1b	20,000
2.	Field labor costs	a. Staff hours	2,000
		b. Hourly cost, $/hour	25
		c. Annual cost, $/year, Line 2a × 2b	50,000
3.	Materials costs	a. Unit cost per participant	10
		b. Number of participants/year	1,000
		c. Annual cost, $/year Line 3a × 3b	10,000
4.	Total service area population		100,000
5.	Targeted population	Percent	10
6.	Targeted population (line 4 × 5)	Number customers contacted (assume 10% positive participation response)	10,000
7.	Publicity costs	a. Marketing cost, $/year	5,000
		b. Advertising costs, $/year	10,000
		c. Annual cost, $/year, Line 7a + 7b	15,000
8.	Evaluation and follow-up costs	a. Labor & consultant, $/year	5,000
		b. Annual cost, $/year, Line 8a	5,000
9.	Total costs (Line 1c + 2c + 3c + 7c + 8b)		100,000
10.	Program cost sharing (e.g., 25% grant funds, partnerships with wastewater, stormwater, or neighboring water utilities) (Line 9 × cost share percentage)	Cost share from other organizations (assume 25%)	25,000
11.	Net agency annual cost, $/year (Line 9 – Line 10)		75,000

and maintenance expenses; and delayed, downsized, or eliminated capital facilities. In this example the need for a new reservoir can be avoided. The volume of water that is not needed also affects the wastewater volume and, therefore, the cost of collection and treatment. The variable chemical and energy cost savings are added to the total.

Step 4 – Benefit–Cost Ratio Calculation (Table 4-5)

The benefit–cost ratio is calculated using the equation as shown below.

$$\text{Benefit–cost ratio} \; \frac{\text{sum of benefits (\$)}}{\text{sum of costs (\$)}} \tag{4-15}$$

The benefit–cost ratio is 1.4 for this example, and the program can therefore be justified on a benefit–cost evaluation basis. Note that the cost share improves the cost-effectiveness and turns a marginal program into a more cost-effective one (Table 4-5).

Table 4-4 Summary of Utility Benefits Calculation

		Value (Example)
1.	Next source of water	Our new reservoir
2.	Average annual (discounted) avoided supply acquisition cost, \$/MG	0
3.	Average annual (discounted) avoided water treatment and distribution costs, \$/MG	1,250
4.	Average annual (discounted) avoided wastewater capacity costs, \$/MG[a]	500
5.	Avoided variable treatment and distribution costs (water plus wastewater if measure reduces both, otherwise just water costs)	
5a.	Chemical costs	
	1. Total annual chemical costs, \$/yr	500,000
	2. Annual costs for chemicals not related to water production, \$/yr	200,000
	3. Avoided chemical costs, \$/yr, Line 5a.1 – Line 5a.2	300,000
	4. Total annual treated water produced, MG (18.5 mgd × 365)	6,750
	5. Unit cost of chemicals, \$/MG, Line 5a.3 / Line 5a.4	44
5b.	Energy costs	
	1. Total annual energy costs, \$/yr	2,000,000
	2. Annual fixed costs for energy, \$/yr	1,000,000
	3. Annual energy costs not related to water production, \$/yr	500,000
	4. Avoided energy costs, \$/yr, Line 5b.1 – Line 5b.2 – Line 5b.3	500,000
	5. Total annual treated water produced, MG (18.5 mgd × 365)	6,750
	6. Unit cost of energy, \$/MG, Line 5b.4 / Line 5b.5	74
6.	Avoided unit variable treatment and distribution costs, \$/MG, Line 5a.5 +Line 5b.6	118
7.	Total average annual unit supply and treatment benefits, \$/MG, Line 2 + Line 3 + Line 4 + Line 6	1,868

[a] Assumes water utility and wastewater utility benefits can be additive because they are part of same government entity in this example

Note: Lines 2, 3, 4 discounted and converted to equivalent annual cost.

Step 5 (Optional) – Net Present Value of Benefits and Costs

If the project or conservation measure water savings are predicted to last for a long period time after the device is installed or to have a large dollar investment over time (i.e., significant annual costs), it can warrant calculation of the net present value of benefits and costs.

This is the basic definition of net present value:

$$\text{Net present value} = \text{present value of benefits} - \text{present value of costs} \quad \text{(This can also be written as NPV = PV benefits – PV costs)} \tag{4-16}$$

Net present value calculations estimate the value of an investment over a course of years or the lifetime of an installed technology. The general form of the present value equation is:

$$PV = \sum n/(1 + r)^{t - t_0} \tag{4-17}$$

Where:

PV = present value

n = value of the measure in year t — i.e., the annual benefit or the annual cost in year t

$r =$ discount rate

$t =$ year

In Microsoft Excel, the formula is:

PV = year 1 value + NPV (rate [r], value year 2, value year 3,...value year t). For an example using Excel, see example in Table 4-6. (4-18)

A net present value analysis can run from very simple to very complex, typically depending on the number of inputs used to derive n in the previous equation.

Discount rate or the real interest rate is the cost of borrowed money minus inflation. The typical discount rate for water utilities is 3 to 4 percent. The example uses a discount rate of 3 percent.

Using data from the earlier example with a water savings of 30,000 gallons per day per year and an avoided cost of $3,200/MG, we can calculate the present value of the benefits (PV benefits):

$$\text{Present value benefits (PV benefits)} = \text{Value of items in year } n \times \text{PV factor}$$
$$\text{PV factor} = 1 / (1 + r)n \qquad (4\text{-}19)$$

Equation 4-17 can be used to evaluate the present values of the costs. Written out for each year the equation below will yield the values in Table 4-7.

$$\text{Present value costs (PV costs)} = \text{Cost of items in year } n \times \text{PV factor}$$

Then the net present value is the difference of the total values in Tables 4-6 and 4-7, i.e., the net present value (NPV) = present value of the benefits – present value of the costs.

Using the data from our example:

NPV = $1,409,933 – $658,958 = $750,975

The goal is a positive NPV, which means that the benefits of the projects are higher than the costs of the project.

This method can be used by a water utility to evaluate residential or commercial projects or for large infrastructure. Many different types of projects can be calculated using

Table 4-5 Summary of Utility Benefit–Cost Ratio Calculation

		Value (Example)
1.	Costs	
	a. Total annual costs, $/yr (costs summary in step 2)	100,000
	b. Cost share from others (such as wastewater utility), $/yr	25,000
	c. Total program (net) costs, $/yr (Line 1a – Line1b)	75,000
2.	Benefits	
	a. Unit water supply and wastewater benefits, $/MG (benefits summary in step 3)	1,868
	b. Lifetime water savings, MG (water savings calculation in step 1)	54.75
	c. Total utility benefits, $/yr, Line 2a × Line 2b	102,273
Benefit–cost ratio, Line 2c / Line 1c		1.4
Benefit–cost ratio is greater than 1.0. Program design for this measure is cost-effective.		

Table 4-6 Calculation of Present Value of Benefits

Year	Savings Multiplier	Annual Water Savings, gpd	Days per Year	$/MG	*t*	PV factor	PV
1	0	0	365	$3,200	0	1	$0
2	1	30,000	365	$3,200	1	1.03	$34,019
3	2	60,000	365	$3,200	2	1.0609	$66,057
4	3	90,000	365	$3,200	3	1.092727	$96,200
5	4	120,000	365	$3,200	4	1.1255088	$124,530
6	5	150,000	365	$3,200	5	1.1592741	$151,129
7	5	150,000	365	$3,200	6	1.1940523	$146,727
8	5	150,000	365	$3,200	7	1.2298739	$142,454
9	5	150,000	365	$3,200	8	1.2667701	$138,304
10	5	150,000	365	$3,200	9	1.3047732	$134,276
11	5	150,000	365	$3,200	10	1.3439164	$130,365
12	4	120,000	365	$3,200	11	1.3842339	$101,255
13	3	90,000	365	$3,200	12	1.4257609	$73,729
14	2	60,000	365	$3,200	13	1.4685337	$47,721
15	1	30,000	365	$3,200	14	1.5125897	$23,166
						Present value of the benefits =	$1,409,933

Table 4-7 Calculation of Present Value of Costs

Year	Annual Costs	*t*	PV Factor	PV
1	$75,000	0	1	$75,000
2	$75,000	1	1.03	$72,816
3	$75,000	2	1.0609	$70,695
4	$75,000	3	1.092727	$68,636
5	$75,000	4	1.12550881	$66,637
6	$75,000	5	1.159274074	$64,696
7	$75,000	6	1.194052297	$62,811
8	$75,000	7	1.229873865	$60,982
9	$75,000	8	1.266770081	$59,206
10	$75,000	9	1.304773184	$57,481
			Present value of the costs =	$658,958

benefit–cost ratios or net present values such as new equipment in laboratories (Figure 4-8) or large number of toilet replacements in a hotel (Figure 4-9).

CONDUCTING BENEFIT–COST ANALYSIS OR NET PRESENT VALUE FOR MULTIPLE CONSERVATION MEASURES

When analyzing multiple measures together over a long time period, a water utility can elect to use software to accomplish the calculations. The following example is provided below that shows how a water utility can group measures and then conduct a water savings analysis for a couple of different levels of water savings as well as budgets. In Table 4-8, 23 measures were analyzed first individually for cost-effectiveness. The measures are

Figure 4-8 Commercial water surveys of large laboratory buildings often involve evaluating the making and use of distilled water

Source: Maddaus Water Management

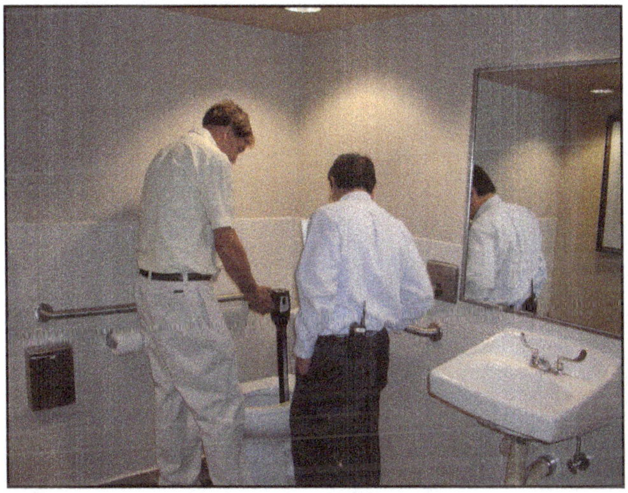

Figure 4-9 Testing flush volume of commercial toilets may reveal that they are out of adjustment and use more water than designed

Source: Maddaus Water Management

sorted by utility; commercial, institutional, and industrial (CII); landscape; and residential categories.

As shown in Figure 4-10, the example has three separate programs – A, B, and C:

- Program A: "Existing Program" option includes measures that the water utility currently offers.
- Program B: "Optimized Program" includes individual measures that were selected by the City. This program includes all the measures included in Program A plus six additional measures: an AMI measure, two CII targeted measures, a school education measure, a recycled water conversion incentive measure, and an agricultural water audit measure.
- Program C: "All Measures Analyzed" presents a scenario where all 23 measures are implemented.

Table 4-8 Conservation Measures Evaluated Example

Utility Measures	CII Measures	Landscape Measures	Residential Measures
Water loss	High-efficiency faucet aerator, showerhead, and soil moisture sensor giveaway	Residential outdoor water surveys	Single-family indoor water surveys
Advanced metering infrastructure (AMI)	Top water users program (top customers from each customer category)	Large-landscape outdoor water audit	High-efficiency faucet aerator, showerhead, and soil moisture sensor giveaway
Pricing	CII rebates to replace inefficient equipment	Large-landscape water budgeting/monitoring	Hot water on demand pump systems rebate
Public information	Require plan review for new CII	Financial incentives for irrigation and landscape upgrades	Residential clothes washer rebate
School education	Promote high-efficiency pre-rinse spray valves	Require weather-adjusting smart irrigation controllers and/or rain sensors in new development	Incentive for recycled-water conversions
	Incentive for recycled-water conversions	Rotating sprinkler nozzle rebates	
		Provide rain barrel incentive	
		Water efficient landscape ordinance	

Program Scenarios			
Measures	Program A	Program B	Program C
Water Loss	☑	☑	☑
AMI	☐	☑	☑
Pricing	☐	☐	☑
Single Family Indoor Water Surveys	☑	☑	☑
High Efficiency Faucet Aerator, Showerhead, and Soil Moisture Sensor Giveaway	☑	☑	☑
Hot Water on Demand Pump Systems Rebate	☐	☐	☑
Residential Clothes Washer Rebate	☑	☑	☑
Residential Outdoor Water Surveys	☑	☑	☑
Large Landscape Outdoor Water Audit	☑	☑	☑
Large Landscape Water Budgeting/Monitoring	☑	☑	☑
Financial Incentives for Irrigation and Landscape Upgrades	☐	☐	☑
Require Weather Adjusting Smart Irrigation Controllers and / or Rain Sensors in New Development	☐	☐	☑
Rotating Sprinkler Nozzle Rebates	☑	☑	☑
Provide Rain Barrel Incentive	☑	☑	☑
Top Water Users Program (Top customers from each customer category)	☑	☑	☑
CII Rebates to Replace Inefficient Equipment	☑	☑	☑
Require Plan Review for New CII	☐	☑	☑
Promote High Efficiency Pre-Rinse Spray Valves	☐	☑	☑
Public Information	☑	☑	☑
School Education	☐	☑	☑
Incentive for Recycled Water Conversions	☐	☑	☑
Ag Water Audit Program	☐	☑	☑
California Model Water Efficient Landscape Ordinance	☑	☑	☑

Figure 4-10 Example conservation measure program scenarios

Source: Maddaus Water Management and City of Oceanside

Table 4-9 shows the annual water savings for plumbing code savings only, as well as plumbing code savings with programs A, B, and C implementation in five-year increments. The benefit-to-cost ratio for each conservation program from the perspective of the

Table 4-9 Water Demand Program Savings Projections

Conservation Program Water Savings (acre-ft/year)	2020	2025	2030	2035	2040	Water Utility Benefit-to-Cost Ratio	Community Benefit-to-Cost Ratio
Plumbing code	730	1,527	2,251	2,738	3,113	N/A	N/A
Program A with plumbing code	1,600	2,802	3,348	3,738	4,137	3.66	1.38
Program B with plumbing code	1,643	3,091	4,414	4,811	5,217	5.09	1.93
Program C with plumbing code	1,867	3,408	4,792	5,265	5,673	5.45	2.07

Note: N/A = not applicable

Table 4-10 Economic Analysis of Alternative Programs

	2040 Indoor Water Savings (acre-ft/ year)	2040 Outdoor Water Savings (acre-ft/ year)	2040 Total Water Savings (acre-ft/ year)	Present Value of Utility Water Savings (Benefits) ($)	Present Value of Utility Costs ($)	Net Present Value of Utility (PV Benefits – PV Costs) ($)	Cost of Utility Savings per Unit Volume ($/acre-ft)[a]
Program A with plumbing code	3,200	936	4,137	$24,739,374	$6,755,664	$17,983,710	$267
Program B with plumbing code	3,944	1,272	5,217	$39,844,227	$7,825,857	$25,028,528	$191
Program C with plumbing code	3,960	1,712	5,673	$47,261,257	$8,665,155	$27,170,817	$177

[a] Utility cost of water saved per unit volume ($/acre-ft) = present value (PV) of utility costs over 25 years divided by the 25-year water savings. This value is compared to the utility's avoided cost of water as one indicator of the cost-effectiveness of conservation efforts. It should be noted that the utility cost of water saved per unit volume somewhat undervalues the cost of savings because program costs are discounted to present value and the water benefit is not

water utility and the combined perspective of the water utility plus the customers (called the community perspective) is also presented.

Table 4-10 shows the year 2040 indoor and outdoor water savings for the three conservation programs modeled; the present value of water savings and the present value of costs to the utility and community are also displayed. The cost of utility savings per unit volume of water is shown in the far right column.

REFERENCES

American Water Works Association Research Foundation (Awwarf). 1996. *Impacts of Demand Reductions on Water Utilities.* Denver, CO: Awwarf.

American Water Works Association Research Foundation (Awwarf). 2001. *Commercial and Institutional End Uses of Water.* Denver, CO: Awwarf.

California Urban Water Conservation Council (CUWCC). 2005. BMP *Costs & Savings Study: A Guide to the Data and Methods for Cost Effectiveness Analysis of Urban Water Conservation Best Management Practices.* Draft revision. Sacramento: CUWCC

DeOreo, W.B.; P.W. Mayer; E.M. Opitz; B. Dziegielewski; J.C. Kiefer; W.Y. Davis; and J.O. Nelson. 2016. *Residential End Uses of Water, Version 2 – 4309*. Denver, CO: AWWA Research Foundation.

Maddaus, W.; and M. Maddaus. 2014-2016. Benefit–Cost Analysis Workshop, *2014 Sustainable Water Management Conference Proceedings*. Denver, Colo. CD-ROM

Maddaus, L.; W. Maddaus; and M. Maddaus. 2013. *Preparing Urban Water Efficiency Plans, A Best Practice Guide*. London: International Water Association (IWA) Publishing.

Pekelney, David; Thomas Chesnutt; and W. Michael Hanemann. 1996. *Guidelines to Conduct Cost-Effectiveness Analysis of Best Management Practices for Urban Water Conservation*. Prepared for the California Urban Water Conservation Council (CUWCC). Sacramento: CUWCC. http://infohouse.p2ric.org/ref/36/35595.pdf. Accessed February 21, 2017

US Environmental Protection Agency (USEPA). 2017. "Pre-Rinse Spray Valves." WaterSense, https://www.epa.gov/watersense/pre-rinse-spray-valves. Accessed February 21, 2017.

Vickers, Amy. 2001. *Handbook of Water Use and Conservation: Homes, Landscapes, Businesses, Industries, Farms*. Amherst, MA: WaterPlow Press.

Chapter 5

Creating a Formal Water Conservation Program Plan

This chapter describes the process to develop a utility water conservation plan. It begins by reviewing program goals and continues using the cost-effectiveness techniques described in chapter 4 to establish and evaluate alternative scenarios consisting of a suite or group of measures. Also, a comparison of alternatives is made. An example outline for a water conservation plan is provided.

MEETING THE AWWA/ANSI G480 STANDARD

Each utility will need to review and decide on its ability to design and fund a conservation program on a level to meet the G480 Water Conservation Program Operation and Management standard (AWWA 2013). This is a voluntary standard that can be adopted by water providers at their own discretion. The G480 standard describes the critical elements of an effective water conservation program and encompasses activities undertaken by a utility within its own operations to improve water use on the supply side through distribution system management and on the demand side through customer billing and education practices. The G480 standard includes the following:

- Dedicated staff for conservation efforts (point of contact)
- Conservation planning
- Integrated resources planning
- Public information and education
- Water waste ordinance
- Universal metering practices
- Nonpromotional water rate

- Monthly or bimonthly billing based on metered use
- Landscape efficiency program
- Water loss control program

Additional requirements and verification components are described in detail in the standard document itself. AWWA's vision for the G-series of voluntary management standards is to enhance public and regulatory confidence while demonstrating continuous improvement efforts of the water industry as a whole. G-series standards offer water utilities of all types and sizes outcome-oriented self-assessment tools that go above and beyond established regulations and whose performance is verified by peer evaluators. Through the G-series effort, AWWA is seeking to foster excellence in water utilities through self-evaluation and peer evaluation. The goal of the standard is to encourage continuous improvement, thus providing the utility with confidence that it can increase consumer satisfaction with utility services. An outline of the standard in the form of a detailed checklist is provided in appendix B.

REVIEW OVERALL PROGRAM GOALS

As discussed in chapter 2, before embarking on selecting measures to be included in a plan, it is important to revisit and refresh the programmatic goals established at the outset of the project. Do any of these goals and objectives for the program lend themselves to a particular design of a group of measures? Is the only goal to a meet a specific water savings target or a level of investment (annual budget) or state or local regulatory requirements? Is there consideration of meeting the AWWA/ANSI G480 standard? What other goals are useful to consider in program design? Listed here are a few examples of program goals that could influence the program design:

- Maintaining existing conservation program (setting a baseline for continuing the status quo and what programs savings and costs are projected near term and long term)
- Complying with regulatory requirements (e.g., water loss control program)
- Educating customers and the public
- Providing the most cost-effective incentives to customers
- Achieving a specific water savings target with the least cost (e.g., target mandating a certain number of gallons per capita per day in a future year).
- Achieving water savings regardless of or with limited consideration for costs (scarcity case)
- Reducing peak demand (maximum hour or maximum day) to assist with future supply goals
- Saving energy (perhaps in partnership with energy utility)
- Reducing wastewater flows
- Reducing stormwater flows to reduce water quality impairment of local streams, rivers, and beaches due to pesticides, insecticides, and fertilizers from excessive dry-weather irrigation runoff
- Enforcing compliance with building codes and other mandates

As described further in this chapter, scenario planning can be accomplished by setting up a few programs to compare and contrast new programs compared to the current one. Having multiple program scenarios (taken as packages of individual conservation

Figure 5-1 High-efficiency toilet installed on college campus
Source: Maddaus Water Management

measures) to choose from can be a useful approach to help managers, elected officials, and the community engage in the dialogue. The goal of grouping measures into scenarios is to understand the different costs and resulting water saving option for future investment in water efficiency.

The following figures illustrate some of the potential conservation measures. Figure 5-1 pictures a high-efficiency toilet that has been installed on a college campus as part of the university's conservation program. Figure 5-2 pictures artificial turf used for soccer fields, which creates a year-round playing field with significantly reduced water usage. Figure 5-3 presents a water smart irrigation controller which, when properly installed, has been proven to save water over the use of time clocks. Figure 5-4 exhibits a rain sensor that is now required on new automatic irrigation systems in Monterey, California.

EVALUATING BENEFIT–COST OF POTENTIAL MEASURES

Chapter 4 showed how to evaluate the benefits and costs of individual water conservation measures. In this chapter, the focus is on analysis of how much water the measures would save, how much they would cost, and what the benefit–cost ratios are if the measures were run on a stand-alone basis – that is, without interaction or overlap from other measures that might affect the same end use(s). Note that measures with benefit–cost ratios less than 1.0 have a negative net utility benefit.

Most water conservation programs consist of multiple water conservation measures. Therefore, it is customary to combine measures into alternative program scenarios and reanalyze the benefits and costs for the alternative programs. When deciding which measures to include in an alternative scenario, the following factors are normally considered:

- **Water savings over the planning period:** These can be averaged over 30 years, or water savings calculated for a particular year, such as 2030, can be the focus.

Figure 5-2 Artificial turf used on a soccer field
Source: Maddaus Water Management

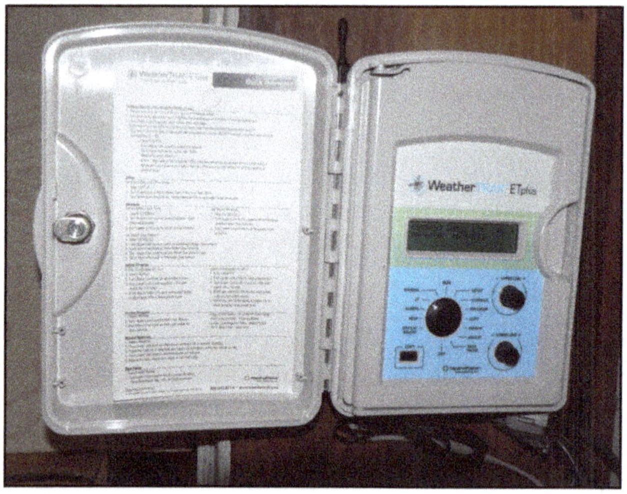

Figure 5-3 Water smart irrigation controller
Source: Maddaus Water Management

- **Utility benefits and costs:** Those benefits and costs that the utility would receive or spend.
- **Community benefits and costs:** Community benefits equal utility benefits plus customer energy (cost to heat water) benefits. Community costs include utility and customer costs.
- **Water benefits:** These are based on the avoided cost of new treated water expressed as dollars per million gallons ($/MG) and avoided wastewater costs (wastewater benefits) also expressed as $/MG dry-weather flow reduced.
- **Required initial year(s) budget for the utility:** Costs can include measure setup, annual administration, and payment of rebates or purchase of devices or services as specified in the measure design.
- **Customer costs:** These costs may affect the measure's acceptance and normally include costs of implementing the measure and maintaining its effectiveness over its useful life.

Figure 5-4 Rain sensor from an automatic irrigation system
Source: Maddaus Water Management

In developing a water conservation plan, it is helpful to develop three to four alternatives of increasing effectiveness, which could be named, for example, program A, B, and C. Here is an example array of program scenarios:

- Program A could include current efforts. For example, it could include three measures: public education, water losses reduction (where needed), and residential retrofit.
- Program B normally includes program A measures plus additional measures as an increase to reach a water savings target at some future date (e.g., 2030 at the end of the planning period). It may be designed to be the midpoint between program A and C, and generally consists of 5–15 measures, all relatively cost-effective. The measures are less aggressive than those in program C, yet are still able to save significant amounts of water.
- Program C often includes 10–30 measures. It normally includes program A measures and usually all of program B measures, plus additional measures. Measures that either saved a small amount of water or were not cost-effective (benefit–cost ratio less than 1.0 and a high cost of water saved) are usually eliminated. Aggressive mandatory (regulatory) measures are often included in this program.

In this way, the three programs represent the range of options from a current (or low-impact modest increase in effort) program to a very aggressive program. Although this can be done with three programs, in some cases four to five alternative scenarios may be needed to represent a full range of choices. Also, some measures may have two levels of implementation, with one being less and one more intensive.

Sometimes the water utility management team and/or a conservation advisory committee may suggest an additional program alternative for menus of measures, after reviewing the results of the evaluation of the first three. More description of stakeholder involvement in the planning process is provided in appendix C.

Alternative Program Evaluation

Table 5-1 presents an example of key evaluation statistics for the water service area. Assuming all measures are successfully implemented, projected water savings and

Table 5-1 Results of an Example Alternative Scenarios Evaluation

Conservation Alternative Scenarios	Water Utility Benefit–Cost Ratio	2030 Water Savings (mgd)	2030 External Water Savings (mgd)	Present Value of Water Utility Costs ($1,000)	Cost of Water Saved ($/MG)
A	1.95	7.65	3.52	62,600	860
B	2.35	14.53	7.77	93,400	720
C	2.50	19.59	10.56	117,900	690

Note: Present value is determined using an interest rate of 3%. Cost of water saved is present value of water utility cost divided by total 30-year water savings.

Source: Levin et al. 2006

wastewater reductions for 2030 in million gallons per day are shown, as are the costs of achieving this reduction. The costs can be expressed in three ways. Shown in the table are the total present value over the 30-year period, the money utilities would need to budget in the first five years to get the program under way, and the cost of water saved. Figure 5-5 is an illustration of the relationship between the water savings at the end of the planning period (in this case 2040) versus the present value of costs to achieve those savings.

Water savings are cumulative in that program A is plotted at the program A plus plumbing code savings, B is A plus the water savings from B, and so on. This figure shows an example where the marginal returns first increase and then decrease as more money is spent to achieve savings. The cost of water saved decreases from program A to program B and increases with program C. The benefit–cost ratios would increase from program A to B, and the ratio for C is a little lower than for B (in this example all measures are cost-effective). The increasing effectiveness in moving from program A to B is because of the inclusion of relatively low cost but effective measures in program B. Program C has more aggressive and costly measures and so the slope of the curve begins to decrease. In this example, one would be inclined to select program B as the most cost-effective. It can also be said that program B represents the point of diminishing returns; that is, further investments in conservation beyond program B achieve further savings, but at a higher unit cost ($/MG). In some communities, it may make sense to select a program that is beyond local cost-effectiveness due to water scarcity issues or regulatory requirements. (For example, state targets for gallons per capita per day are effectively an unfunded state mandate that must be paid by the local community for the greater benefit for managing watershed benefits beyond the local service area boundary.)

PLAN SELECTION CRITERIA

At this stage in the process, the plan is a program alternative that is merely a group of conservation measures. Later, the plan will include a budget, staffing, and implementation requirements and schedule. Before this additional effort is expended, the alternative program scenarios should be narrowed to a preferred option or recommended plan.

The job of water conservation planners is to present information needed by decision makers who will select the final water conservation plan. In some cases, the planner will be asked for a recommendation; in other cases, only an evaluation of alternatives and suggested plan selection criteria will be requested. Focusing on an appropriate set of plan selection criteria is important whether making a decision or just presenting technical and nontechnical information concerning alternative plans. There is no one list of criteria that fits all situations. Presented next are criteria that the planner may deem appropriate for the local situation.

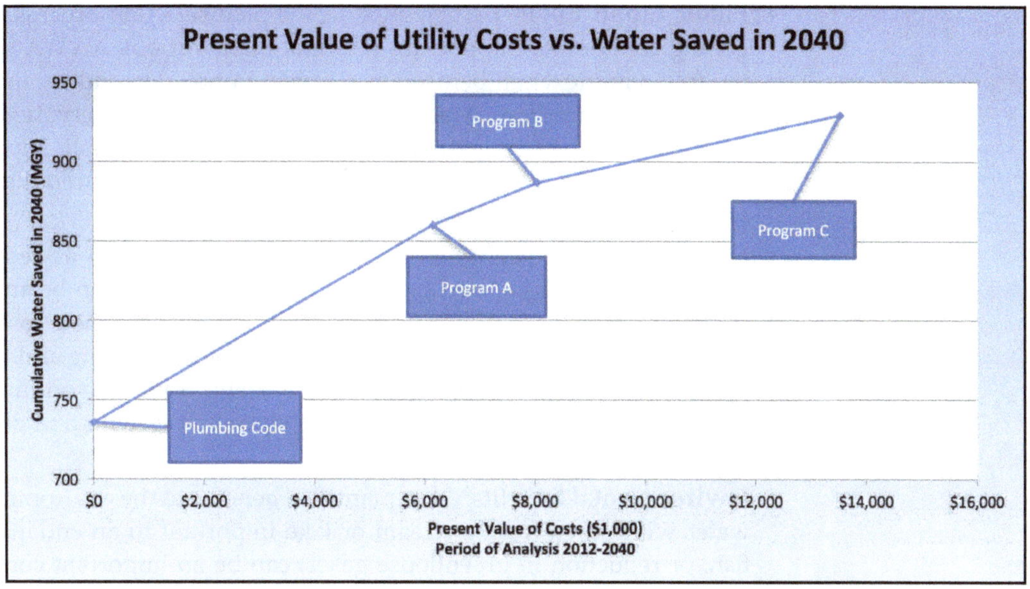

Figure 5-5 Example of the relationship between present value of utility costs and water saved in a future year

Source: Maddaus Water Management

- **Likelihood of meeting goals:** How well does the alternative plan meet goals (such as a water savings target in a specified year)? Goals were discussed in chapter 2. How well the plan will meet those goals should be documented and quantified.
- **Cost-effectiveness:** Alternative plans can be evaluated for cost-effectiveness using such measures as the benefit–cost ratio or the cost of water saved in dollars per million gallons ($/MG). The higher the benefit–cost ratio or the lower the cost of water saved, the more cost-effective is the alternative. Small differences in these parameters should be ignored because they must be prepared with estimated water savings and costs. The plan should seek to illustrate the best value for the conservation dollars expended.
- **Effect on long-term versus short-term supply situation:** In some cases, the plan may obviate the need to expand a water supply or water treatment plant. If the plan can do this in the short term, it will produce more immediate benefit for the agency than deferring a capital project that is not needed for 10 to 20 years. In some cases, a conservation program can stretch an existing supply until a more permanent long-term supply can be developed. Or in the case where no new supplies are available, a conservation program can help maximize existing remaining supplies.
- **Regulatory requirements:** A number of states require that water conservation plans be prepared, often by a certain date. Such requirements generally specify the contents of the plan; with few exceptions the level of aggressiveness of the plan – i.e., the amount of water that must be saved – is left to the discretion of the local water agency. In some cases, the measures that must be evaluated are specified, such as customer metering, leak reduction, and public education. In a few cases regulatory agencies have set targets, such as a per capita use reduction target, say 20 percent by 2020. Where these occur, meeting the target often takes precedence over having a program that is cost-effective. Instead, the goal is to meet the target at the minimum utility cost.

- **Public input:** Public participation in the plan selection process is important for getting community support for the final plan. Various AWWA publications and the following text can be used as a guide on how to conduct an effective public participation program. Many utilities have a water conservation advisory committee that provides at least part of the needed process. Public support for the plan is important and when choices are involved, it can be ensured that the public will weigh in on which alternative is preferred.
- **Energy benefits:** Quantified benefits to the utility (such as reducing pumping costs, as well as associated greenhouse gas emissions) can be an important consideration. AWWA has numerous publications that can help to quantify energy savings through improved management. Reduced pumping and treatment due to water conservation is most often a very valuable finding to optimize on-peak and off-peak demands (loads) through water treatment systems and pumping to water storage facilities.
- **Environmental benefits:** Nonquantified benefits to the environment (such as less water withdrawn from a stream or lake important to an endangered species of fish, or reduction in greenhouse gases) can be an important consideration. This type of information may be found in reports and environmental studies by the water utility and others when expansion of a source of water is being considered. At a minimum, the lessened environmental impact correlating with a more aggressive conservation plan can be described.
- **Partnership opportunities:** Some conservation measures offer more of a potential for partnering with other funding agencies. Energy companies will often co-sponsor rebate programs for measures that save hot water, such as efficient clothes washers. Wastewater agencies sometimes will co-sponsor measures that reduce wastewater flow, such as low-flush toilets or efficient commercial process equipment. In other instances, in-kind services can be arranged or cost sharing of a regional public education program can be established among local water agencies. Pooling of resources may allow access to more effective and more expensive means of reaching the public, such as paid television advertisements.
- **Budget and staffing requirements:** No matter how cost-effective a program is, if the utility cannot afford it, it is unlikely to be implemented. Often, realizing the program benefits requires up-front financing of the program to save the amount of water needed to defer the capital project at a future date. Many water agencies are reluctant to ramp up conservation program funding for a promised future benefit. Other factors to consider include the following: Who will do the work? Will in-house staff supplemented by contractors do the additional work, or will new staff be hired? How difficult is it to hire new staff? Rectifying budget and staff imbalances requires skillful marketing of agency decision makers by conservation staff.
- **Community impacts:** Depending on the extent of the public participation program, the community impacts may be identified during the process, and mitigation strategies developed. In other cases, the community impacts can and should be assessed early on. How will the community react to restrictions on where and how much turf can be planted? Will the community object to bans on nonrecycling car washes (which may raise the price of car washes), or decorative fountains, or other water uses? In these situations, public opinion surveys are an effective means of gauging public support or resistance to specific measures.

- **Other criteria:** The process should be open, and the public and other stakeholders should be encouraged to submit their own list of criteria for plan selection. This could include other benefits and costs not included in the evaluation. In some cases, the agency board of directors will be very explicit in this regard; in others, they will look to staff and consultants to provide direction through the process.

SELECT THE RECOMMENDED PLAN

There are several steps involved in selecting a recommended plan:

1. Finalize the list of selection criteria
2. Develop and use a public participation program
3. Rate the alternatives
4. Make a formal selection

The recommendation may be made at the staff level and then reviewed by management for buy-in on the selected program. An additional step could include review and comment by stakeholders or an advisory committee. Careful attention to detail and addressing comments received while formulating the recommendation are important to achieving formal recognition of the plan for implementation.

It is important that a recommended plan with its numerical results is preferably presented with enough time for a thorough review and comment period. It is most helpful to have the recommended plan formally approved by a governing board, advisory committee, or the highest level of management before the written draft plan document is finished. This step of getting buy-in on results has been found to be more time efficient than rewriting chapters of the final plan with multiple draft versions distributed for review.

Create a Matrix

An example evaluation matrix that planners may find useful is shown in Table 5-2. The general approach is to compare alternatives. This can be done with short word statements, a plus or minus system, or ratings on a scale of 1 to 5 where 5 is the highest rating. In the example shown, program B is superior because it has the most total points.

Staff can do the rating, or they can use a water conservation advisory committee to rate some of the alternatives against some or all of the criteria. This information can be presented along with the recommended program alternative to aid decision makers having buy-in on the selected plan.

CONSERVATION PLAN CONTENTS

There are many drivers for a having conservation plan, and there may be flexibility in its content and organization. There are also water conservation plans that are (or should be) required for compliance with federal and state financing requirements. For example, states may require that a water use efficiency plan be in place to access the Clean Water Act state revolving loan fund (SRF) and the Safe Drinking Water Act SRF. On June 10, 2014, the Water Resources Reform and Development Act (WRRDA) of 2014 was signed into law by the president. Title V of WRRDA contains the Water Infrastructure Finance and Innovation Act (WIFIA) that also requires a water conservation plan be adopted and implemented to access funding. The state requirements to access these funds may drive the necessary content and/or organization of the plan. As a result, water conservation

Table 5-2 Example of a Rating Matrix for Comparing Alternative Conservative Plans

	Comparison or Rating		
Criteria	Program A	Program B	Program C
Meet goals	3	4	5
Cost-effectiveness	5	4	3
Supply benefits	3	4	5
Regulatory	5	5	5
Public support	4	5	2
Environmental benefits	3	4	5
Partnership opportunities	5	5	3
Budget	5	4	3
Staffing	3	4	5
Community impact	4	5	5
Other (specify)			
TOTAL	40	44	41

planners should check any necessary state requirements before embarking on developing the detailed contents of the plan and ideally would have reviewed these requirements at the outset of the planning effort (see Section 2).

The state of California has one of the most detailed requirements for a water management plan (of which water conservation is an integral component) in its Urban Water Management Planning Act, which is applicable to any water utility serving more than 3,000 connections or 3,000 acre-feet of potable water supply. California Water Code Sections 10608–10656 covers division 6, part 2.55 "Sustainable Water Use and Demand Reduction," which includes gallons per capita per day reduction targets, and part 2.6 "Urban Water Management Planning," which outlines the overall requirements for plans submitted to the California Department of Water Resources (CDWR). The Urban Water Management Plan is the key document to be adopted by local utilities every five years that provides a comprehensive summary of information and a water balance of current and future water supplies and demands projected for the next 20 years. CDWR offers guidance for preparing plans and performs an "acceptable review" of the plans for completeness. Plans need to be complete and accepted in order for a utility to be eligible for any state loans or grants for any water related project (that must be listed in its Integrated Water Resource Management Plans for water supply, wastewater, stormwater, flood control, etc.).

The following shows the contents of a typical plan. Figure 5-6 is an example table of contents for a water conservation plan report.

I. Service Area, Climate, and Demographics
   Describe service area and climate of service area, including current geographic boundaries of service area, current and projected population (five-year intervals projected out 20 years, the suggested time frame for a plan), and other demographic factors affecting water planning.

II. Source of Water
   A. Identify and quantify existing and planned sources of water. Describe the water treatment that is or will be required to make the water potable.

III. Supply Vulnerability
   A. Describe reliability of water supply and vulnerability to seasonal and climatic shortage, examining an average water year, a single dry year, and multiple dry years. Complete this section only if section VII is also included in the plan.
   B. For any water supply not available at a consistent level of use (taking into account specific legal, environmental, water quality, or climatic factors), describe plans to replace that source with alternative sources or water demand management measures.

IV. Water Transfers/Exchanges
   A. Describe opportunities for exchanges or transfers of water on a short-term or long-term basis.

V. Existing and Projected Water Use
   A. Quantify past and current water use (five-year intervals).
   B. Project future water use (five-year intervals projected out 20–30 years) by water use sectors or customer classes, if data is available from the water billing system. Suggested sectors could be single-family, multifamily, commercial, industrial, institutional/government, landscape, sales to other agencies, saline water intrusion barrier use/groundwater recharge/conjunctive use (or any combination thereof), and agricultural use.
   C. Project the timing of the need for more water supply and additional water treatment capacity, in the absence of additional water conservation.

VI. Demand Management Measures
   A. Provide a description of each water demand management measure that is currently implemented or is scheduled for implementation detailing the dates that the program went into effect, the affected customer classes, and the number of affected units (persons or accounts). The plan will describe the steps necessary to implement any proposed measures, including, but not limited to, the following suggested list of conservation measures (also called best management practices):
      - Water waste prevention ordinances
      - Metering
      - Conservation pricing
      - Public education and outreach
      - Programs to assess and manage distribution system real loss
      - Water conservation program coordination and staffing support
      - Other demand management measures that have a significant impact on water use as measured in gallons per capita per day, including innovative measures, if implemented
   B. A schedule of implementation for all measures proposed or described in the plan.
   C. A description of the methods, if any, that the water supplier will use to evaluate the effectiveness of measures implemented.
   D. An estimate, if available, of existing conservation savings within the supplier's service area and effect of such savings on the supplier's ability to further reduce demand.

E. An evaluation (which could take into account each of the following items) for each measure listed that is not currently being implemented or scheduled for implementation:
    1. Economic and noneconomic factors, including environmental, social, health, customer impact, and technological factors
    2. Benefit–cost analysis, identifying total benefits and total costs
    3. Which combinations of measures meet any applicable targets of other regulatory requirements
    4. Description of funding available to implement any planned water supply project that would provide water at a higher unit cost
    5. Description of the water supplier's legal authority to implement the measure and efforts to work with other relevant agencies to ensure the implementation of the measure and to share the cost of the implementation. The objective of the evaluation is that the utility must demonstrate the reasons for not implementing specific conservation programs.

VII. Detailed water shortage contingency analysis and drought/emergency action plan (including draft implementing ordinance). Drought/emergency plans are useful, and most states that now require conservation plans also require a drought plan as an element of the long-term plan.

VIII. Detailed information on recycled water and its potential, including coordination with local sewer service agency.

IX. Recommended long-term conservation plan.

A. Include a summary of the measures selected and their implementation schedule.

B. Project the required budget and staffing over the life of the plan.

C. Discuss the monitoring and reporting that will be done to ensure that the water conservation goals are being met.

D. Provide a resolution or other evidence of official plan adoption by the utility containing a statement of intent to implement the plan.

FORMAL ADOPTION OF THE PLAN

The need for formal adoption varies by jurisdiction and purpose of the plan. Some regulators require the utility to formally adopt the plan before the plan is considered official and is used to meet other obligations, such as water rights permits, to qualify for state grant and loan programs and so on. Once adopted, it is often filed with the state water agency, which oversees utility planning requirements. In some cases, it may be appropriate for the plan to be an internal plan for the utility, while in others, it may require local or state government approval. Nonetheless, any plan should obtain whatever approval is appropriate before it is implemented.

The public should be engaged in the development or adoption of a plan. A detailed description of public involvement approaches in the design and adoption of the plan is presented in chapter 6 and appendix C. The goal of the public participation program is to inform the public, seek community input, and ultimately obtain approval of the plan developed by water agency decision makers. The process should involve at minimum one public meeting so that all stakeholders receive a chance to offer input.

The Madison, Wisconsin, case study in appendix A gives a good overview of a plan and the process used for a medium-size city.

Section 1 Introduction and Summary
Purpose and Scope of Plan
Plan Submittal Requirements
Plan Development and Public Participation
Plan Elements
Resolution for Adopting the Plan
Section 2 Study Area Characteristics
History of Water System
Demographic Forecasts
Section 3 Analysis of Historical and Projected Water Demand
Historical Water Use
Analysis of Water Use by Customer Group
Summary of Historical and Projected Demand, Without Conservation
Impact of New Plumbing Code on Water Use
Impact of Building Codes on Water Use
Section 4 Water Supply
Sources of Water
Groundwater
Surface Water
Overall Supply and Demand Balance
Section 5 Recycled Water Plan
Results of Previous Studies
Plans for Reuse
Section 6 Current Water Conservation Program
Measures Implemented by Water Wholesaler and the Utility (i.e., City)
Management of Water Loss Control Program
Section 7 Alternative Water Conservation Measures
List of Conservation Measures Considered
Section 8 Evaluation of Long-Term Water Conservation Measures
Menu of Water Conservation Alternative Programs
Estimated Water Savings
Costs of Measures
Results of Benefit-Cost Analysis
Section 9 Recommended Plan
Selection Criteria
Description of Recommended Plan
Projected Water Savings
Benefits
Implementation Schedule
Budget and Staffing
Section 10 Water Shortage Plan
Worst-Case Water Supply
Plan Elements
Water Use Restrictions
Water Supply Emergency
Water Rate Structure in Shortage
Appendix A State Water Management Planning Requirements
Appendix B Detailed Description of Selected Water Conservation Measures
Appendix C Benefit-Cost Model Output
Appendix D Drought Shortage Ordinances

Figure 5-6 Example water conservation plan report table of contents

REFERENCES

American National Standards Institute (ANSI)/American Water Works Association (AWWA). 2013. Water Conservation Program Operation and Management Standard G480-13. Denver, CO: AWWA. http://www.awwa.org/store/productdetail.aspx?productid=35009354, accessed February 24, 2017.

Levin, E.R.; W.O. Maddaus; N.M. Sandkulla; and H. Pohl. 2006. "Forecasting Wholesale Demand and Conservation Savings." *Journal AWWA*, 98:2.

This page intentionally blank.

Chapter 6

Stakeholder Involvement

Involving stakeholders in the design of water conservation programs can help ensure that the program has the best opportunity to be a successful and sustainable conservation program. In light of increasing investments in water conservation programs, there is a vital need to educate and involve the public in planning and decision making, and that process should be supported by dedicated staff and a budget. This chapter can assist in developing an understanding of public participation and the important part it can play in developing publicly supported programs and building a group of community advocates and partners for conservation. Two case studies in appendix A (Cary, North Carolina, and Sacramento, California) give examples of involving stakeholders in the planning and implementation phase. For more detailed guidance on developing an in-depth public involvement program, see appendix C.

INVOLVE THE PUBLIC IN SUPPORT OF THE PROGRAM

To build and maintain grassroots support for the program, it is important to involve key stakeholders in developing and implementing the plan. As the cost of water conservation programs for many larger systems (i.e., those serving more than 100,000 persons) is now reaching into the millions of dollars, there is an ever-increasing need for public – and specifically ratepayer – involvement in the design and adoption process of the conservation plan. This chapter includes some general guidelines for conducting an effective program that will involve the public and build support for the programs the utility wishes to carry out.

The planning and decision-making process should be open and should invite participation of affected stakeholders. The public participation program can vary from a public hearing at a single water supplier board meeting to a full-scale public involvement program. The former may be all that is required to secure conservation plan adoption, but it is unlikely to develop the type of support necessary for the program to be successful.

Why Involve The Public In Water Efficiency Planning Process?

- **More buy-in:** When the conclusions have been vetted through an open, transparent, equitable, and inclusive process, it reflects the broad diversity of views within the community or region.
- **More educated action:** Citizens are apt to take action when water users are involved in the solution and understand how to save water in the community.
- **More successful implementation:** Citizens are more engaged in the implementation phase of the project when they have the opportunity to buy in during the planning phase of the project.
- **More enthusiastic support:** Professionals in sectors like landscape, institutional, commercial, and industrial are more enthusiastic about and involved with planning and implementation of the program.
- **More funding approval by decision-makers:** With more understanding by local leadership, more funding support can be gained for the programs.

Sometimes public participation is mistakenly confused with marketing a utility program. The article "Don't Confuse Marketing with Public Participation" (Katz 2002) addresses this issue.

SUPPORTING AN ENGAGED STAKEHOLDER PROCESS

This section provides guidance on the following topics in support of an engaged stakeholder process during the conservation planning process:

- Identify target audiences
- Address inequities in terms of range of viewpoints
- Select forums for consensus building
- Leverage tools for consensus among stakeholders
- Conduct a successful process

As stated previously, more in-depth information, including how to develop a public involvement work plan, is provided in appendix C.

Identify Target Audiences

Every planning process will have a unique list of target audiences that should be involved. The following list identifies participants to be considered:

- The general public
- Elected officials
- Utility managers
- Community leaders
- Environmental groups
- Economic development and business organizations
- Wholesale water customers
- Local and regional agencies
- State and federal regulatory agencies

Figure 6-1 Design center for low-water-use plants
Source: Maddaus Water Management

- Developers
- Neighborhood and community associations
- Large water users
- The media

Through contacts and interviews, this list can be used to identify stakeholders and specific group representatives that should be invited to participate.

Shown in Figure 6-1 is a kiosk demonstrating options for a new development model home. It is provided through a partnership between a developer (Shapell Homes) and a utility (East Bay Municipal Utility District in northern California) and is designed to assist new homeowners with landscape decisions.

Address Inequities

When deciding which target groups to focus on, those who might be impacted by proposed conservation measures should be considered. Ask "Who will feel left out of the process or feel unfairly treated?" For example, if a rebate program is targeted to only single-family homes, apartment manager associations should be contacted so they can be apprised of why this decision was made. If clothes washer rebates are targeted to single-family homeowners, only then should other ways be found to communicate with and involve apartment residents. Finally, if commercial programs are targeted to high water users, small business owners should be approached in other ways.

Forums for Consensus Building

Once a utility has established its goals and identified the appropriate stakeholders, it needs to bring them together to discuss priorities in order to develop the final plan. It is important to determine the right forums, tools, and processes to follow that will help stakeholders come to agreement on the right path to follow.

Figure 6-2 Water efficiency training begins in the classroom using a slide show to explain principles in a workshop environment

Source: Maddaus Water Management

1. **Public meetings:** Although they are difficult to arrange, they provide a very open, informal, and participatory forum that can enhance the utility's relationship with its customers. They may be too unstructured for consensus building.
2. **Citizens advisory committees:** These committees allow a broad range of stakeholder input on a regular basis over the duration of the project.
3. **Workshops:** These can provide a participatory process for exchanging ideas and information. By coming together in a workshop, interested parties can focus on specific issues and concerns and build consensus (see Figure 6-2).
4. **Task forces:** These are more formalized and exclusive groups that are usually charged with the task of devising and recommending solutions to specific problems.

Tools for Consensus Building

In addition to the procedural models listed above, there is a wide variety of methods that can be used in conjunction with and to support the selected model. These methods include

1. **Participant surveys:** These can provide a knowledge basis by cataloging participants' basic concerns and expectations, and can be used to highlight divergent views on key issues that will need to be resolved. These surveys can be random telephone surveys of the general public or one-on-one interviews with key stakeholders.
2. **Issue or discussion papers:** These help define the issues and provide a common basis of knowledge about an issue or set of issues. While they do not attempt to resolve issues, they can be the catalyst for educating and starting discussions with stakeholders.

Figure 6-3 Irrigation restrictions during a recent drought in Florida
Source: Maddaus Water Management

3. **Position papers:** Members of groups can state broad positions on policy issues that can be helpful in finding common ground among groups with diverse perspectives.
4. **Policy statements:** These go further by committing participants to a specific position. Draft statements are circulated until consensus on a final version can be reached. They reinforce the outcome of the consensus-building process.
5. **Memorandum of understanding (MOU):** These are particularly appropriate for regulatory agencies needing to clarify roles and improve coordination. By signing an MOU, different groups can commit to a process to resolve a specific issue or agree to take unilateral action in support of the consensus.

In some situations, consensus must be built during trying times, such as a severe drought (Figure 6-3).

Guidelines for Conducting a Successful Process

Key strategies that can help the planner create and implement a successful stakeholder process leading to conservation plan adoption can be summarized as follows:

1. Identify obstacles to plan preparation early on and take steps to mitigate them.
2. Involve the public to get their input, in plan development, to build community support for the conservation programs, and to better reflect and include community-supported goals for program implementation.
3. Use the conservation plan process to set priorities, justify an adequate budget, and maximize benefits at the lowest cost (fund programs that are cost-effective to the utility).

4. Communicate with key stakeholders inside and outside the utility to ensure smooth implementation, including adequate staff and budget.
5. Use rates as a conservation tool and factor conservation into the rate-setting process to avoid reduced-revenue surprises.
6. Use public participation to garner public support for the plan and its programs.
7. Find supporters and practitioners of conservation among stakeholders to give input on programs, and help implement them in the community.
8. Publicize plan successes as a way to ensure a continuous flow of management support and budget.

In addition to maintaining an honest and open participatory atmosphere, the following are some general tips to help ensure success:

- **Set realistic goals:** Organizers and participants must bring realistic goals to the table. This process will not solve all problems. Stay focused on the key issues that require consensus.
- **Optimize participation:** Only the stakeholders that need to be involved should be involved. Smaller groups of 25 or fewer can be more efficient in making decisions and reaching consensus.
- **Discourage hidden agendas:** Some groups or associations may attempt to use the stakeholder process to advance goals unrelated to water conservation and efficiency. Directly addressing such efforts is usually sufficient to dissuade repeated attempts.
- **Create access and openness:** Many potential participants will expect the process to be open and resources, such as key staff, available during and between meetings. Information germane to the process should not be privileged. Participants should not feel inhibited by the nature of the proceedings.
- **Allow due process:** All affected parties have the right to be heard.
- **Sense the need for closure:** It is important to know when to close the process because of diminishing returns. It may be necessary to settle for consensus on a limited number of issues and resolve the remainder in a more traditional forum. In some cases, when competitive solutions are available, consensus may not be desirable. In other words, it is acceptable to disagree on certain issues and to clearly state positions so that ultimate decision makers can appreciate the different points of view as they try to balance water issues.
- **Publicize successes:** Publicize how much water the community, by using these techniques, is saving on an annual basis.

Public participation (Figure 6-4), in particular consensus building, is one of the key features of water conservation planning and program implementation. Public input can result in a better plan. If the public is involved from the beginning, the support for new programs, including the staffing and budget, is more likely to be forthcoming.

A wide variety of consensus-building techniques exists. Once the first seven steps have been conducted (see sidebar, "Eleven Steps to Engaging the Public") and the stakeholders identified, the nature of the specific techniques can be selected.

Figure 6-4 Vendor fair for landscape contractors featured efficient weather-based irrigation technologies

Source: Seattle Utilities

Eleven Steps to Engaging the Public

The American Water Works Association Research Foundation has previously developed helpful handbooks including "Public Involvement – Making It Work" (AwwaRF and AWWA 2001), which describes 11 steps to engaging the public in any water resources project including selecting a conservation program plan strategy. This step-by-step process outlined below to engaging stakeholders in water use efficiency planning efforts has been specifically tailored from water infrastructure projects.

Most of the steps are related to the research needed to form the foundations of an effective program. The 11 steps are listed below, the first eight of which are related to developing the strategy:

1. **Framing the problem:** Focus on issues for why the water efficiency program needs to succeed. Be specific regarding goals and objectives for public involvement.
2. **Identify the constraints:** Determine the issues that can be negotiated with the public and those that cannot, such as regulatory or political mandates and spending limits.
3. **Identify and describe decisions and planning milestones:** Develop a schedule that shows when the public will be able to provide input into decisions.
4. **Identify potential participants:** Determine potentially affected participants, or stakeholders, who need to be involved.
5. **Determine vulnerability and "must resolve" issues:** Focus efforts on those issues and groups that are likely to generate any conflict and prioritize as to those with the greatest need for resolution and those that may proceed without resolution.
6. **Determine the appropriate level of public involvement:** Establish what level of involvement is needed in order to engage stakeholders.

7. **Select techniques:** By completing the first six steps, the planner can save time and money by selecting the most suitable public participation from many available techniques (see Tools for Consensus Building earlier in chapter). Focus on the right tool for where the input is needed in the planning process.
8. **Finalize a public involvement strategy:** The planners should detail a schedule, budget and staffing requirements for carrying out the strategy designed.
9. **Implement and monitor the strategy:** Periodic monitoring is necessary to ensure that:
   (a) The time frame of the problem has not changed (i.e., continuing to plan for long-range water efficiency program implementation in the midst of a drought crisis)
   (b) The issues remain valid and progress is continuing in an acceptable time frame (9–12 months is typical to fully engage stakeholders and complete the technical planning process)
   (c) The techniques being used are effective and stakeholders remain engaged
10. **Develop a consensus-based statement of goals:** To finalize this strategy session, the planners and stakeholders should issue a statement, based on the previous steps, that defines the agreed-upon mission for the planning process, along with specific goals and objectives.
11. **Manage change:** The process must be flexible enough to adapt to changes in schedule, political environment, staff, or critical issues.

SUMMARY OF RECOMMENDATIONS

The top six key points from this chapter that can help the planner involve the public to create and implement a successful conservation plan can be summarized as follows:

1. Early on, identify water issues in the community that may need to be dealt with first before the role of water conservation in resource management can be discussed, and take steps to mitigate them.
2. Involve the public to get their input on plan development, to build community support for the conservation program, and to better reflect and include community-supported goals for program implementation.
3. Communicate with key stakeholders inside and outside the utility to ensure smooth implementation, including adequate staff and budget.
4. Use public participation to garner public support for the plan and its programs.
5. Publicize plan successes as a way to ensure a continuous flow of management support and budget.
6. Build success by partnering with practitioners of conservation among the broader community.

REFERENCES

AwwaRF and AWWA. 2001. "Public Involvement – Making It Work: Highlights Report." Denver, CO: AwwaRF and AWWA.

Katz, Sara. 2002. "Don't Confuse Marketing with Public Participation." *Journal AWWA*, 94:7 (38–39). www.awwa.org/Portals/0/files/resources/publicaffairs/pdfs/ConfuseMktgWithPublicComm.pdf, accessed February 24, 2017.

Chapter 7

Implementing the Plan, Managing Revenue Impacts, and Evaluating Performance

This chapter describes the implementation and evaluation of the ongoing efforts related to a utility conservation program and plan. It begins by following the development and adoption of a formal water conservation plan, as described in chapter 5. The implementation issues that are covered include setting a budget, scheduling staff, partnering, and training. Successful implementation also requires that conservation managers work with other parts of the utility to manage revenue impacts that could result from decreases in demand. This chapter also presents some methods for identifying, tracking, and evaluating meaningful and practical program performance and water-saving measures.

IMPLEMENTATION CONSIDERATIONS

Budget

The plan should lead to a budget request for support of plan activities. This can involve confirming the need for existing staff and possibly hiring additional staff to run programs or contracting out certain programs to private companies. The latter approach is often preferred when the plan requires the use of specialists for a limited amount of time. In addition, adding staff is often a long and difficult process.

Table 7-1 Example of a Detailed Budget for the First Two Years of Implementation

Plan Element	Staff Labor, $	Contract Labor	Materials and Publicity	Incentives (Rebates)	Total
Public education	10,000	5,000	10,000		25,000
Fixture rebate	25,000	25,000	5,000	45,000	100,000
Residential water audits	20,000	40,000	10,000	5,000	75,000
Totals	55,000	70,000	25,000	50,000	200,000

The budget could be formatted as shown in Table 7-1 with particular details depending on the local agency needs.

Prioritizing

Budget priorities may need to be established for the following reasons:

- Only some of the funds requested were approved;
- Preferences stated by agency management;
- Staff limitations;
- Need to hire staff and/or contractors;
- Time of the year or other sequencing considerations for measures; or
- Cash flow constraints.

Schedule

A generalized schedule for recommended program strategy is helpful to align with a cost-effectiveness analysis and also for budget planning. An example of a strategic schedule for years that conservation measures are planned to start and stop is presented in Figure 7-1.

When the budget is confirmed, more detailed schedules should be developed and tracked. An example of a plan element implementation approach, which should be created with milestones and deadlines as part of the conservation program's annual work plan, is presented in Figure 7-2. The schedule provides the priorities and available budget for focusing staff on what should be done and when.

PROGRAM MONITORING, EVALUATION, AND PERFORMANCE MEASUREMENT

A water conservation program should be dynamic. Changes should be expected based on how well the program meets the goals developed in the water conservation plan. Observations concerning the water supply situation, growth projections, customer participation and satisfaction, and water savings achieved should be made. These observations should be used in evaluating the programs and in improving them to ensure that they remain effective and economical.

Defining Tracking Measures, Including Key Performance Indicators

The City of Sacramento Department of Utilities (DOU) offers an example of a water conservation program that uses tracking metrics and performance indicators. From 2013 to 2015, a seven-member DOU committee met monthly and developed a measurement, tracking, and reporting system for 61 key performance indicators (KPIs) that included water conservation measures. In 2013, the DOU committee determined that 66 percent of the

61 measures could be tracked and reported on, while the remaining needed additional work before they could be accurately reported. The committee saw improvement in 2014, finding that 85 percent of the measures could be tracked and reported on; only 15 percent needed additional work.

No.	Measure	Time Period	2015	2016	2017	2018	2019	2020	2021	2022	2023	2024	2025	2026	2027	2028	2029	2030	2031	2032	2033	2034	2035
1	System Water Loss Reduction	2015-2035																					
2	Advanced Metering Infrastructure	2021-2035																					
3	Large Landscape Budget-Based Water Rates	2018-2020																					
4	General Public Information	2015-2035																					
5	Public Information (Home Water Use Report)	2018-2035																					
6	Residential Leak Assistance	2018-2035																					
7	Single Family Residential Surveys	2015-2035																					
8	Plumbing Fixture Giveaway/Opt	2015-2017																					
9	Residential Ultra High Efficieny Toilet Rebates	2015-2020																					
10	High Efficiency Clothes Washer Rebates	2015-2026																					
11	High Efficiency Clothes Washer - New Development	2021-2035																					
12	Hot Water On Demand - New Development	2021-2035																					
13	Toilet Retrofit at Time of Sale	2015-2019																					
14	CII MF Common Laundry Room High Efficiency Clothes Washer	2015-2024																					
15	CII Incentives	2021-2026																					
16	Pre-Rinse Spray Nozzle Installation	2015-2016																					
17	CII Surveys	2021-2026																					
18	High Efficiency Urinal Program	2015-2018																					
19	Public Restroom Fauce Retrofit - MUN	2021-2023																					
20	Public Restroom Fauce Retrofit - COM	2021-2030																					
21	School Retrofit	2021-2030																					
22	Water Efficient Landscape Ordinance	2015-2035																					
23	Single Family Residential Turf Removal	2015-2035																					
24	Multifamily Residential/CII Turf Removal	2015-2035																					
25	Expand Large Landscape Survey/Water Budgets	2018-2035																					
26	Sprinkler Nozzle Rebates	2018-2035																					
27	Gray Water Retrofit	2015-2035																					
28	Residential Rain Barrels	2015-2035																					
29	Climate Appropriate Landscaping and Rainwater Infiltration	2015-2035																					
30SF	SF Conservation Pricing - Water and Sewer	2018-2035																					
30MF	MF Conservation Pricing - Water and Sewer	2018-2035																					
30COM	COM Conservation Pricing - Water and Sewer	2018-2035																					
31	Single Family Multifamily Dishwasher Rebates	2018-2022																					
32	Hot Water Recirculation Systems	2018-2022																					
33	Rewarding Businesses for Adopting Best Practices	2020-2035																					
34	Additional Building Code Requirements for New Development	2018-2035																					
35	Innovation Incubator Program	2021-2035																					

Note: CII = Commercial, Industrial and Institutional
MUN = municipal
COM = commercial
SF = single family
MF = multifamily

Figure 7-1 Example of a recommended 20-year program plan implementation strategy

Source: City of Santa Cruz. Water Master Plan. January 2017.

Plan Element	Activity	January	February	March	April	May	June	July	August
Public education	Design	■							
	Create materials		■						
	Media event			■					
	Booth at county fair						■		
	Etc.								
Fixture retrofit	Selective incentive amount		■						
	Hire contractor			■					
	Marketing				■	■			
	Kickoff event						■		
	Process rebates							■	■

Figure 7-2 Example of a plan element implementation schedule for first year of implementation

Source: City of Sacramento Department of Utilities, 2015.

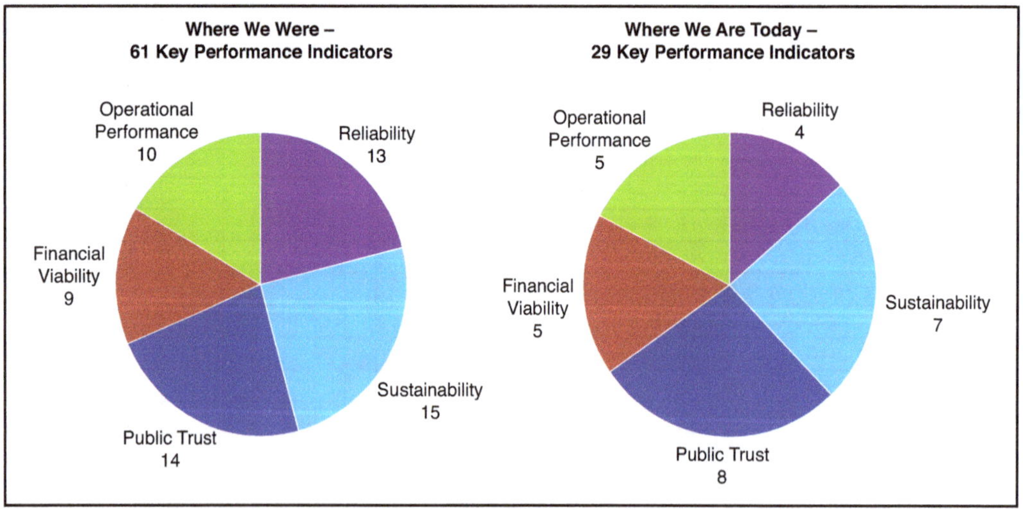

Figure 7-3 Continual improvement management program results by strategic goal

Source: City of Sacramento Department of Utilities, 2015.

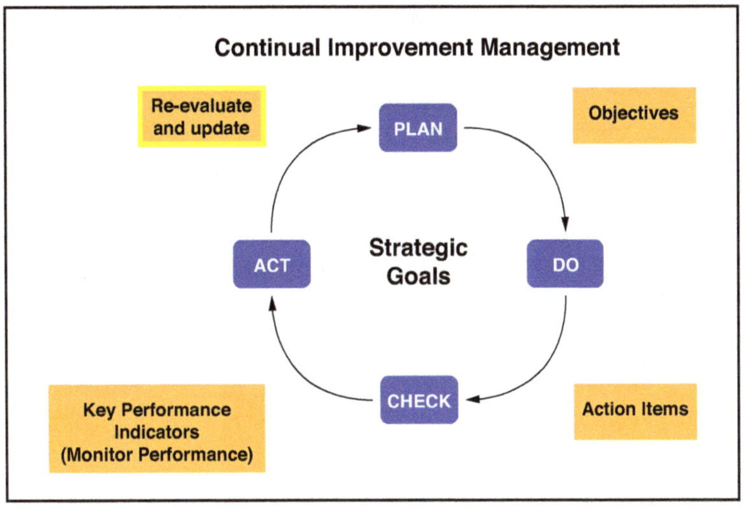

Figure 7-4 Continual improvement management program using key performance indicators

Source: City of Sacramento Department of Utilities, 2015.

The following list and Figure 7-3 present an example of possible sustainability and water conservation goals and key objectives. This example from the Sacramento DOU was set for 2015.

- Validate internal performance measures and results;
- Provide regulators, governance, and stakeholders with confidence in performance; and
- Identify and support performance improvement initiatives and funding justifications.

More than 60 DOU staff were involved, and their roles included champions, such as subject matter experts, responsible parties, and data collectors. The DOU leadership team became the champions for improvement and provided resilient, day-to-day management continuity and direction. Their effective leadership ensured the DOU's direction was understood, embraced, and followed throughout the management cycle. The champions did the following:

- Educated themselves to become knowledgeable about their conservation measures;
- Oversaw their measures and improved performance;
- Met with the committee and data collectors;
- Made recommendations to improve performance; and
- Pulled in resources.

Most importantly, the DOU champions, who were chosen by the division managers and approved by the director, had full support that included training, if needed, support for managing the measures, and top management's commitment to implementing change to improve performance. Figure 7-4 shows the continual improvement management program's goals and the approach implemented by the DOU committee.

Leveraging Measurement and Evaluation for Effective Program Management

Measurement is critical to effective management. As a measurement and evaluation team focuses on effective program management, it may be difficult to define useful metrics that are feasible to track and will lead to future program enhancements.

Challenges can be expected, such as deciding what to measure, identifying meaningful measures, and making sure that data are collected in a way that allows meaningful comparisons to be made for later evaluations. These evaluations may include statistical analyses that require preplanning during setup of the measure implementation. Overcoming these obstacles will allow managers to make real improvements, communicate accurate information, and ensure that the information is interpreted correctly when seeking to improve programs in the future.

With established results, efficiencies, and benchmarks, utility staff can show solid results and information based on demonstrated measurement and management. Staff may find it goes a long way in giving decision makers confidence in their performance—aiding them in pursuing resources and providing funding justifications.

In examining opportunities, the utility should look at all methods for encouraging water efficiency, including rebates, free materials or plumbing fixtures, conservation-oriented rate structures, and identification of the public relations value to the customer through recognition programs.

Many utilities are beginning to use behavior modification strategies, such as home water use reporting software, to achieve additional savings and increase participation in incentive programs. By providing customers with home water use reports, the utility is able to leverage social norms, that is, compare the customer's water use with that of similar customers, as well as provide feedback to the customer on water- and cost-saving ideas and available rebates. In addition, goal setting by the customer enables the utility to determine whether this customer engagement will be more effective in changing water use behavior.

Benchmarks for water use can also be used to more easily identify above-normal water use where potential water savings can be found. These benchmarks should include normalizing factors such as average or median water use (gallons) per person and occupancy or living/working/use area (square footage) for each customer group selected. Once the benchmarks have been established, the volume of potential savings and corresponding number of customer accounts can be used to identify the most cost-effective conservation strategies for each customer group. High water use, either by an individual customer or by a customer group, does not by itself indicate inefficiency. However, it does indicate where potential water savings may be found or where limited resources or incentive programs should be focused. Code requirements and their effective dates are also used to identify grandfathered facilities where incentive programs could be used, as opposed to newer facilities that are subject to water efficiency requirements for which code compliance monitoring would be the more appropriate strategy.

Typically, the following four types of program follow-up need to be performed:

1. The utility must keep good records of the impact that the conservation program is having (that is, to measure water savings). Accurate and complete water use data and customer classification systems before, during, and after implementation of a measure are essential to evaluating water savings.
2. The utility should monitor how well the program is performing and whether it is achieving its goals, workload measures (or outcomes), conservation measures (for example, unit–cost ratio, cost of water saved), and effectiveness (outcomes). (See the Sacramento case study in Appendix A and the Sacramento Water Conservation performance measures worksheet [Figure 7-5]).

Table 7-2 New Water Savings Achieved in 2013 (mgd)

	New Long-Term Customer Savings, mgd						Transitory Savings, mgd			
	1% Conservation				Seattle Low Income					
	Hardware	Behavior	Rates	Code	Seattle Low Income	Subotal	Curtailment & Economy	Water System	Subtotal	Total
Water System	–	–	–	–	–	–	–	2.0	2.0	
Residential indoor	0.29	0.1	0.3	0.5	0.3	1.5	0.7		0.7	2.2
Residential landscape	<0.01	0.3	0.1	–	–	0.4	0.6		0.6	1.0
Commercial domestic	0.04	0.2	0	0.3	–	0.5	0.9		0.9	1.4
Commercial process	0.15	0.1	0.1	–	–	0.4	1.6		1.6	2.0
Commercial landscape	0.01	<0.1	<0.1	–	–	0.0	0.2		0.2	0.2
2013 Total	0.5	0.7	0.5	0.8	0.3	2.8	4.0	2.0	6.0	8.8

Note: – = not applicable

Source: Seattle Saving Water Partnership 2013

3. The utility must monitor the actual program costs and compare them to the planned costs in order to evaluate cost-effectiveness.
4. The utility should monitor how well various plan elements are being received by customers (AWWA 1993). For example, see Table 7-2 to review Seattle's Saving Water Partnership conservation accomplishments report (Saving Water Partnership 2013). The Partnership separated program element permanent savings from temporary savings, including drought curtailment programs and the 2009 recession, which were expected to rebound to normal levels in time.

The following steps can be taken when implementing careful, effective measurement management:

1. *Establish a performance measurement/management committee:* This committee should include representatives from each section to communicate and coordinate specific tasks. Representatives can come from many places within the organization (such as water resources, operations, or planning/business), and each should be notified of the goals and project plans when recruited.
2. *Identify and record department levels of service:* The purpose of this step is to discuss the levels of service with your committee in order to develop and implement measurement, tracking, and reporting systems. This can be done by developing strategies and schedules and by coordinating central reporting systems.
3. *Assign subject matter experts (SMEs):* Within your committee, assign SMEs to specific KPIs. SMEs should be assigned by taking into account performance measure content and the representative's job function. Following the assignment of SMEs, experts need to receive information on what their roles and responsibilities are in regard to the program.
4. *Perform a follow-up procedure that involves the following:*
   - Review preliminary analysis and receive input;
   - Coordinate with information technology to discuss data collection (for example, shared drive, wiki site, SharePoint, or e-room); and
   - Meet with performance measurement/management committee to discuss strategy, schedule, roles and responsibilities, and SMEs selected and responsible for collecting data (at least two measures per month).

<table>
<tr><td>Strategic Goal</td><td colspan="6">Public trust/sustainability/water conservation</td></tr>
<tr><td>Objective</td><td colspan="6">Review and update performance on targets periodically</td></tr>
<tr><td>Action Item</td><td colspan="6">Develop and implement performance measurement/tracking/reporting system
How the performance measures inform utility staff on what changes must be made; how staff embeds those changes back into the organization; and how staff creates value based upon the data</td></tr>
<tr><td rowspan="4">Designation</td><td>Responsible</td><td colspan="5">Operations and maintenance</td></tr>
<tr><td>Accountable</td><td colspan="5">Operations and maintenance division manager</td></tr>
<tr><td>Consulted</td><td colspan="5">Director and division managers</td></tr>
<tr><td>Informed</td><td colspan="5">All other units</td></tr>
<tr><td colspan="2" rowspan="2">Department Task(s)
(See Task Details following this figure)</td><td colspan="2">Resources Required</td><td colspan="2">Proposed Timeline</td><td rowspan="2">Dependencies</td></tr>
<tr><td>Materials</td><td>Staff</td><td>Start</td><td>Finish</td></tr>
<tr><td></td><td></td><td></td><td></td><td></td><td></td><td></td></tr>
<tr><td></td><td></td><td></td><td></td><td></td><td></td><td></td></tr>
<tr><td></td><td></td><td></td><td></td><td></td><td></td><td></td></tr>
<tr><td></td><td></td><td></td><td></td><td></td><td></td><td></td></tr>
</table>

Figure 7-5 Example of an action item tracking worksheet

Source: City of Sacramento Department of Utilities, 2015.

A schedule for implementing and evaluating key performance measures with a committee can be shown in an action item tracking worksheet. This worksheet includes tasks, resources required (materials and staff assigned), proposed timeline, and dependencies (see Figure 7-5).

PROGRAM PERFORMANCE TRACKING AND EVALUATION

The following section presents information about tracking and evaluating program performance.

Tracking Actual Costs

The committee or water conservation group should set up an accounting system that allows all costs related to implementation to be separately accounted for. These costs include labor, contracts, materials, and other expenses that are accounted for in the same way the plan elements are budgeted (see Table 7-1). Costs should be accumulated monthly and reported annually.

Progress Versus Planned Schedule

The planned schedule for implementing the plan elements shown, in Figure 7-2, can be used to track progress. Milestones of when activities are actually completed will show where the utility is falling behind and where it is excelling. These trends may indicate that implementation costs are rising above planned levels or that the program does not have adequate staffing. Quarterly tracking of progress will allow adjustments to be made before more serious problems develop.

Monitoring Performance and Customer Feedback

Customer surveys are often used to identify customer motivations and barriers to various conservation measures. Statistically significant customer surveys by mail, telephone, or online are fairly inexpensive, with an average cost of $5,000–$15,000 per survey. Professional survey companies are very good at preparing unambiguous questions that can be answered in a limited amount of time. The maximum time and number of questions varies, but usually there are about 30 questions and it takes 15–20 minutes to answer them. Surveys are a common way to track and monitor progress of public information and education programs for which water savings are difficult to measure.

Most large utilities have completed a number of surveys and will usually share their questions with other utilities. Negative customer feedback indicates that the program needs to be changed or improved. However, positive feedback reinforces that the program is succeeding.

If negative responses are received for any questions, consider revising the program by doing the following:

- Evaluating alternative conservation measures;
- Modifying existing measures to increase participation; and
- Focusing efforts on other potential water-saving ideas.

In addition to customer surveys, it is important for the utility to evaluate its programs and progress constantly and internally. Ask these common questions:

- Are the measures being achieved? If not, why not?
- Is public response positive? If not, why not?
- Are the specific conservation measures contained in the program effective? If not, why not?

Managers can review and monitor performance by discussing what was accomplished (workload measures or outputs) and how efficiently resources were used (conservation measures or unit–cost ratios). They can then relay results that describe how well customers' and internal and external stakeholders' needs were met (effectiveness of measures or outcome).

Tracking Actual Water Savings

One way to track water use and overall water conservation savings is to create a conservation performance model. Such a model can be used to track monthly water use, in gallons per day per account (gpd/a), for each customer class. The model can be used to identify trends in water use and reductions that are a result of water conservation programs or other reasons. An example was given in Chapter 3 (see Figure 3-5). A model can be set up by doing the following:

Figure 7-6 Water audits can indicate potential and achieved commercial water savings, for example, new commercial dishwashers can be very efficient

Source: Maddaus Water Management.

1. Obtain monthly water sales and number of accounts billed for each billing period, for each customer class to be analyzed, going back at least three years, preferably 10 years.
2. Put the data in a preprogrammed spreadsheet model that computes the monthly water use in gpd/a and a moving average trend line of the gpd/a. Separate total gpd/a into indoor and outdoor water use. This is done by analyzing the winter use and calling that indoor use (when it can be assumed there is no landscape watering). Note that this method does not work in areas where landscape is watered in all winter months, such as Florida and the southwestern United States. In those cases, assume an indoor or winter water use based on published studies and assume the rest is landscape use.
3. Update the model with new data on a quarterly or annual basis to track water use.
4. Summarize the results for all customer classes.

Another way to verify water savings in a particular sector is to conduct water audits. This may provide a better understanding of the market penetration of new efficient technology in customer accounts (Figure 7-6).

Using Statistical Methods for Program Evaluation

Existing statistical methods can be reviewed for their applicability to specific program evaluation situations. Methods include several types of regression analysis and use significance testing, such as chi-square, factor analysis, analysis of variance, and engineering estimates.

Analysis of conservation programs can be developed for each method; then the methods can be subjected to a screening process to identify which ones best meet a specified set of criteria. Afterward, the methods are matched up with the following applications, including their subelements:

- Tracking overall monthly conservation performance model by customer class and provider including:
  - Seasonality,
  - Baseline demand,
  - Indoor and outdoor consumption and performance,
  - Weather normalization,
  - Naturally occurring conservation (due to existing laws and codes), and
  - Savings from programmatic conservation measures.
- Analysis of specific conservation programs (compiled or survey data) including:
  - Before and after analysis,
  - Control versus test group analysis,
  - End-use information,
  - Customer attitude surveys, and
  - Use of engineering estimates.
- Peaking analysis.
- Evaluation of drought/restriction periods.
- Measuring response to water price changes (elasticity).

Evaluation of Meeting Program Goals

Program goals can be qualitative or quantitative. For each goal that is adopted, a method of measuring performance should be developed. Progress can be tracked using tables or matrixes so that performance is readily evident. If the goal is not being met, either the plan element should be changed or (if the goal was unrealistic or inappropriate) the goal should be revised. Performance should be compared with goals annually, prior to the utility's annual budget-setting process.

Figure 7-7 is an example worksheet that outlines the KPIs and program workload, efficiency, and effectiveness measures. This worksheet can be instrumental in concentrating staff on the goals, efficiency, and effectiveness of delivering the program and in achieving the intended outcomes. The worksheet program statement and measures should be reviewed with the leadership team and used with other department managers, council members, and external stakeholders as a management tool for assessing program performance and for highlighting service efforts and accomplishments.

ADDITIONAL IMPLEMENTATION OPPORTUNITIES, TRAINING, AND OTHER SUPPORTING TASKS

The following section presents additional information on implementation opportunities, training, and other tasks to support a conservation program.

Agreements and Partnering

Many agencies implement programs that can be based on a regional scale by partnering with other water, wastewater, energy, stormwater, and solid waste recycling agencies. For

Program Name: Water Conservation

FTEs: ____ Responsible Manager: ______________________

Program Statement:

To: Provide dependable high quality, cost-effective water in a fiscally and environmentally sustainable manner and to provide our regulators, governance and stakeholders' confidence in our performance and identify and support performance improvement initiatives and funding justifications.

By: Investing in operating, maintaining, and replacing infrastructure in compliance with regulatory requirements and our long-term financial plan

1	Gallons of water produced/supplied	Yes
2	Gallons of water consumed per account per day, by customer class	Yes
3	Number of participants reached in community outreach events, by outreach program	Yes
4	Number of water conservation audits conducted	Yes
5	Number of participants served through water conservation incentive/rebate programs, by program	Yes
6	Miles of water main surveyed for leaks/number of services surveyed	Yes
7	Number of customer calls answered	Yes
8	Number of operations & maintenance (O&M) paid overtime hours	Yes
9	Number of administrative penalties and notice of violations received	Yes
10	Number of work hours consumed per leak repair	Yes
11	Number of participants reached in community outreach events per FTE	a
12	Number of water conservation audits conducted per FTE	a
13	Cost per water conservation audit conducted	a
14	Gallons of water consumed per capita per day	a
15	Real [water] losses per service connection per day (gallons)[b]	Yes
16	Average speed to answer (ASA) customer calls	Yes
17	Percent change in gallons of water consumed per capita per day	a
18	Percent change in gallons of water consumed per account per year, by customer class	a
19	Ratio of current annual real losses (CARL) to unavoidable annual real losses (UARL) (infrastructure leakage index [ILI] from AWWA M36 [best industry practices])[c]	a
20	Gallons of water conserved through water conservation incentive/rebate programs, by program	a
21	Gallons of water saved from proactive leak detection and repair program	Yes
22	Number of water quality complaints received per 1,000 customers	a
23	Percent reduction in the ratio of paid overtime hours to hours of regular work	Yes

a = can calculate
b = real losses are physical water losses from the system up to the point of customer consumption
c = The ILI is a highly effective performance indicator for comparing (benchmarking) the performance of utilities in operational management of real losses (AWWA 2016).

Figure 7-7 Example worksheet of water conservation performance measures

Source: City of Sacramento Department of Utilities, 2015.

example, a low water use landscape program created to meet water conservation objectives may also help reduce contaminated stormwater flows, thus helping both the water agency and stormwater agency achieve their goals. Coordination with other efficiency programs offered by electric and natural gas utilities can also significantly enhance the effectiveness of water conservation programs by leveraging the resources and common incentive programs of these other utilities.

Each plan element should be reviewed for partnering opportunities. Examples include the following:

- Public information programs with other water and wastewater agencies;
- School education programs with a general environmental theme cosponsored by water and wastewater and solid waste recycling agencies;
- Regional efficient clothes washer rebate programs with other water agencies and energy companies;
- Regional residential and/or commercial toilet rebate programs with wholesale and retail water agencies and wastewater agencies;
- Regional residential water audit programs with other water agencies; and
- Regional landscape audit and efficiency programs with other water and stormwater agencies, parks, and school districts.

When entering into a partnership agreement to provide in-kind services or funding, there should be a clear statement of purpose and enumeration of benefits to the partners. For example, a toilet replacement program should identify the water savings to the water agency and the wastewater flow reduction to the wastewater agency for every toilet replaced. This will allow both the water agency and wastewater agency to report and follow their relative savings for each toilet. In addition, the avoided costs can be calculated and used to establish the offered rebate amount. The water savings and benefit–cost analysis should be used to provide this information.

In any partnership a lead agency should draft and outline the agreement, which is then reviewed by the committee. Due to legal constraints, the committee should not write the final agreement or contract. A memorandum of agreement can be signed to identify responsibilities and commitments.

Adopt Ordinances and Other Tools

Some measures are regulatory in nature, such as toilet retrofit on resale or landscape requirements on new development. For these measures, the utility may need to seek adoption of the requirements by other authorities such as a city or state law. There are many examples of ordinances being written by other agencies to accomplish most regulatory programs. A literature search should be conducted prior to initiating preparation of a program.

Over time, many of the easiest conservation strategies will have been used, and more aggressive, ordinance-based measures may need to be considered if the utility's conservation goals are to be achieved. Because of this, a greater regulatory focus can be expected, one that embeds conservation into new development and deters excessive discretionary water use. Many utilities can anticipate offering fewer small rebates as local and state laws as well as federal manufacturing requirements replace incentive programs, especially in the residential sector. The focus is likely to shift to more comprehensive changes that save larger volumes of water, short-term incentives to promote new technologies, programs that target high water users, and marketing efforts that heighten consumer awareness of water use patterns and choices.

Incentive Programs

In developing incentive programs, the utility should first analyze the benefits of any incentive program, including, but not limited to, reducing the cost of treating and delivering water, potentially delaying the need for additional treatment capacity, and maximizing the life of existing water supplies. Water efficiency programs can be considered as a way to buy back capacity in the system. Obtaining a comparable public benefit from an incentive program is especially important when distributing public funds to private persons or entities.

Factors that may be considered when developing a rebate program include whether the program:

- Is needed to help achieve the utility's quantified or qualitative conservation goals;
- Serves as an incentive to get customers to do what they otherwise would not have done without the rebate, rather than simply subsidizing a business or customer sector;
- Provides comparable value to the utility in terms of avoided costs related to water and wastewater treatment and distribution and development of new supply;
- Is cost-effective to the utility and the customer;
- Gathers needed data on new and innovative technology;
- Facilitates access to limited or hard-to-reach markets; and
- Increases public awareness on the need to conserve.

It is equally important to determine when an incentive program should be phased out. Factors to be considered when determining whether to terminate a rebate program include the following:

- Local codes and ordinances that require the use of the incentivized measure, process, or equipment;
- New federal efficiency standards that affect market availability;
- Evidence of significant "free ridership" so that the rebate no longer incentivizes desired changes in behavior or choices in higher-efficiency equipment; and
- Market saturation of the rebated equipment that exceeds 20 percent of market availability.

Training

Experienced water conservation staff is often hard to find. Even if the staff is experienced, they often need training to be sure they operate effectively within a specific department. Water conservation is not a technically difficult subject, but certain basic business and communication skills are often required, even for technicians who implement programs such as water audits. It is important that all water conservation staff be well trained. There are many resources available through other organizations such as AWWA, Alliance for Water Efficiency (http://www.allianceforwaterefficiency.org/) and the California Water Efficiency Partnership (CalWEP, formerly known as the California Urban Water Conservation Council) that can assist in training.

Many how-to papers on water conservation topics have been published in AWWA conference proceedings (https://www.awwa.org). In addition, AWWA has a considerable amount of information available, including manuals and other types of publications. See the AWWA Water Conservation Resource Community website (https://www.awwa.org/resources-tools/water-knowledge/water-conservation.aspx) for more information.

Some statewide groups and water associations offer training on request. CalWEP offers several-day training courses (https://calwep.org) on water conservation. More specialized training can be found through the Irrigation Association (https://www.irrigation.org), which trains technicians to do landscape water auditing and budgeting. In addition, many larger water agencies with trained staff are willing to help other agencies that lack staff expertise. Figure 7-8 shows a water efficiency training session in Sacramento, California.

Figure 7-8 Water efficiency training session

Source: Maddaus Water Management

MANAGING REVENUE IMPACTS

The implied reduction in revenue to a water supplier is often a controversial obstacle to the adoption and implementation of conservation plans. The impact of the conservation program on revenue must be evaluated from the start, and rates need to be adjusted to ensure stable and sustainable revenues for the water utility. This section provides information on how to address this concern. Raising water rates or changing the water rate structure can be among the most sensitive decisions a water utility makes. These changes often stimulate public debate.

As part of the water conservation planning process, a demand projection with planned conservation is usually generated. Demand reductions during the implementation phase are predictable and usually on the order of 0.5 to 2 percent per year. Water rates are usually revised every three to five years, so this level of demand reduction would amount to as little as 1.5 percent to as much as 5 percent of total system water use, depending on the level of conservation being planned and the rate-setting time interval. Also, inflation in utility costs, which is on the same order of magnitude as planned demand reductions, would need to be addressed. Successful use of the public involvement processes described in chapter 6 can go a long way to developing new rates that the public will accept.

Utility rates can be an important tool for encouraging efficiency in customers' use. Depending on rate design and the proportion of revenue generated from water usage as compared to fixed fees, the result of increased customer efficiency can result in reduced revenue, which can lead utility decision makers to be less supportive of conservation programs. A utility's revenue needs and implementation of a conservation program can interact in the following ways:

- When the water rates remain unchanged but an aggressive conservation program is started, there will be revenue impacts. However, as mentioned previously, on an annual basis, they are small.
- Water rates can also be used to encourage conservation. Various rate types send price signals to customers, encouraging them to reduce water use. In this case, lower water sales are an expected outcome of the program and a part of it.

- A utility can use both of the previously mentioned techniques to reduce water use. An aggressive conservation program will include effective programs that reduce water use in a measurable way. An aggressive water rate structure will also encourage water conservation. Following implementation of both, it is challenging to separate the effects and to determine how much of the reduction was due to programs and how much was due to raising the price of water or changing the rate structure to encourage conservation. This is a difficult question, and there are no easy answers. In recent years, many have used econometric models to isolate and study individual items such as price, weather, population, and economic impacts.

Because the consumer price charged for water and the nature of the utility's conservation programs are often intertwined, it is important to understand how to analyze water savings as a result of programs and to price changes separately. The key to managing revenue impacts of a conservation program (lower water sales) is to predict those lower sales and account for them in the rate-setting process. The following will help the planner understand the price impacts.

Types of Rate Structures

Impacts are a function of the type of rate structure the utility uses. Objectives of rate structure design often include basing the rates on the actual cost of service, providing adequate and stable revenues, providing fairness or equitability among customer classes and volume users, water conservation, and ease of implementation and administration. AWWA manuals M1 and M54 (AWWA 2012, AWWA 2004) discuss water rate design in detail and are excellent references to consult before making any changes in existing rates.

Following are guidelines for those who wish to use water rates to create more incentives to reduce water consumption. Note that it is very important to ensure that any proposed water conservation–oriented rate structure follows accepted rate-setting standards and any applicable legal guidelines.

Water rates that encourage conservation. These rates provide a financial incentive for customers to reduce water use, usually by applying a surcharge on peak-month usage or by charging a higher unit rate for water as the number of units used increases. Conservation rates must be fair; it is therefore essential that conservation rates be developed through a public process that ensures acceptance of the purpose and design of the rate structure. It is important that regardless of the conservation rate structure selected, greater influence on customer behavior can be achieved from a combination of pricing with indoor and outdoor conservation programs than from pricing alone. Conservation pricing as part of a broad demand management program is the most logical approach. Rates that promote water conservation include the following:

- A combination of low rates for baseline minimum water quantity (the same fixed charge every billing cycle for the baseline volume) and high volumetric charges for the amount that the customer uses above the baseline volume;
- Inclining tier rates with volume amounts (or blocks) where higher unit charges are triggered at higher levels of use to encourage conservation;
- Seasonal rates or excess-use surcharges;
- Marginal cost pricing;
- Water budget rate structures where tailored allocations are developed for each customer and rates increase as the allocation is used or exceeded; and
- Rates where the majority of revenue is generated by the volumetric rates instead of the fixed rated or service charge.

Rate structures that provide minimal incentive for water conservation. The following rate structures do not encourage water conservation:

- A declining block structure,
- A flat rate structure (a fixed fee regardless of water use), and
- A uniform rate structure (the same unit charge for water regardless of how much is used).

These types of rate structures, especially the first two, offer little incentive for customers to improve water use efficiency. A useful reference for current water rates in place around the United States is the biennial Water and Wastewater Rate Survey (RFC/AWWA 2016).

Price elasticity. When water rates are raised significantly (beyond an inflationary response), water use often declines. The amount of the decline in demand is called *price elasticity,* which is the percentage change in consumption per percentage change in price. Price elasticity is normally expressed as a fractional reduction related to the proportional increase in the rate, that is, an elasticity of −0.1 means that a 1 percent increase in price will stimulate a 0.1 percent decrease in consumption.

It is difficult to predict changes in water use caused by changes in price. Methodologies for assessing the response to rate changes are covered in cited AWWA manuals and should be considered when deciding whether and how to implement water rate changes. It is helpful for planners to have an understanding of price elasticity concepts, because they may greatly influence the revenue generated and thus the utility's financial situation if water rate structures are not applied correctly. This issue is not covered here; more in-depth reference material, such as AWWA's *Principles of Water Rates, Fees, and Charges* (AWWA 2012), should be consulted.

Computer models are available to help utilities simulate the effects of different rate designs on their revenue streams (for example, the free rate model at www.financingsustainablewater.org). Some factors that affect price elasticity (assuming that the utility already has a rate structure with 40 percent or less cost recovery through fixed charges) include the following:

- *Rate levels:* If a utility has low rates (for example, total water bill is a very small percentage of total household income), an increase in the rates from block to block may have a small effect on the water quantity demanded.
- *Availability of alternative water sources:* The demand for water may be more elastic if customers have access to private irrigation wells, reclaimed water, or some other alternative water source.
- *Environmental awareness and attitudes:* Environmentally oriented customers may be more responsive to water conservation initiatives.
- *Time period:* Water demand is more price-elastic over time. Over the long term, customers can adapt to higher water prices by making investments that can lead to more efficient indoor and irrigation water usage.

Revenue Concerns With Different Types of Rate Structures

Water utility finance managers are responsible for providing a steady and predictable revenue stream. As the recession of 2009–2011 demonstrated, when the economy falters, people use less water and water sales decline. The amount of revenue decline varies depending on the type of rate structure currently in place. The least sensitive are flat rate structures where the same monthly charge is levied every billing period regardless of water use. The most sensitive is inclining block rates where the revenue decreases more than the water sales do if users in the highest block use less. The remedy for this is usually to have a reserve fund to handle this situation. The reserve fund can be used to handle the revenue impacts due to weather changes and seasonal fluctuations in the customer base. Weather (dry or wet) can also affect water sales. For example, an atypically cool, wet season can lead to lower outdoor use and lower revenues.

Billing Cycle Considerations

For water rates to form an effective deterrent to wasteful water use, the rate structure should be designed to have an impact on potentially wasteful customer categories. In addition, the deterrent rates must be effectively communicated so that customers can make informed choices on whether or not to use water. There are three important aspects in communicating these rates:

1. The shorter the billing cycle, the more frequent the reminder to customers of the cost of water. Where there is a chronic shortage of water, utilities can move to monthly meter reading and billing. During prolonged drought periods, monthly billing takes on even greater importance. In contrast, with quarterly billing, there is minimum communication with customers, and this type of billing defeats the purpose of conservation rates. For example, a higher bill for the dry season might be sent to customers during the wet season.
2. Use of advanced metering infrastructure (AMI) has provided water agencies the opportunity to post customers' usage online where they can frequently check on their water usage. AMI also gives water agencies the opportunity to more rapidly notify customers when there are sudden changes in usage as might be caused by a leak.
3. The bill sent to water users should clearly show the amount and cost of water used and be separate from wastewater and other charges. It is also useful to provide consumption history and other comparative data. Including the average water use of the neighborhood also provides an effective tool to let customers know when they are using excessive amounts of water as compared to their neighbors.

Integration of Conservation and Rates

In evaluating the relationship between water price and water use, the following should be considered:

- In the case of indoor water use, customers are unlikely to engage and continue with water-saving habits, such as shorter showers, fewer toilet flushes, and larger laundry loads, on the basis of cost alone. However, higher cost may encourage leak repair. The more efficient water-preserving approach is to install fixtures that ensure water savings. It might be argued that higher water costs will prompt the use of efficient fixtures. However, experience with attempts to reduce indoor water use through pricing alone suggests that utility education programs that stress the necessity for water savings and giveaway/rebate programs are more effective. In a two-tier rate structure, there is little latitude for setting a high second-tier rate without reducing the first-tier rate or changing the balance between fixed and volumetric-derived revenue. Consequently, in using a two-tier rate structure, it is difficult to address outside water use even with differentials for dry-season and wet-season use patterns. It is more effective to engage in conservation programs related to landscape design, irrigation methods and practices, and incentives for

Figure 7-9 Recycled water use in landscape irrigation
Source: Maddaus Water Management.

reductions in use, rather than to rely solely on water pricing to obtain water use conservation improvements.

- A rate structure of three or more tiers can provide separate pricing for essential, discretionary, and excessive water usage. The water conservation rate structure should primarily target the excessive and discretionary usage ranges.

A recycled water use system for landscape irrigation used in a new housing development in California is shown in Figure 7-9. Recycled water used for irrigation can significantly reduce the need for expensive potable water.

Building Ongoing Conservation Into Rate Setting

Water rate setting and conservation program implementation are linked, either by design or inadvertently. When rates are used as a part of the process to encourage conservation, it is hard to separate the effects. A tracking model, as described in chapter 5, will show the reductions, regardless of their origin. Because the goal is to reduce water use, it is essential to know how the total reduction is achieved and how the components that contribute to the total are estimated.

A by-product of water use forecasting (chapter 3) and benefit–cost analysis (chapter 4) is the ability to forecast future water savings. The forecast of water use can be translated into a forecast of water revenues. This may be simple with some rate structures and more difficult with multitier rates. Using this forecast of lower water demand, the future rate increases can be designed to accommodate lower sales. When viewed on an annual basis, the lower sales are usually rather small, usually between 0.5 and 1 percent per year, and are probably more predictable than the effects of inflation on costs, which also must be factored into the rate-setting process. Because most utilities review rates about every three years, it becomes evident that revenue losses caused by a conservation program are very manageable.

Alliance for Water Efficiency's Resources to Align Rates, Revenue, and Conservation Programs

Today, water suppliers must encourage much-needed efficiency and conservation programs to protect limited supplies, while also covering the rising costs of water treatment, delivery, and infrastructure upgrades. In addition to thinking about short-term revenue stability, they must also be planning for long-term fiscal sustainability.

New resources are available to help managers navigate these challenges. The Alliance for Water Efficiency's (AWE's) Financing Sustainable Water initiative was created to provide practical information to guide utilities from development through implementation of rate structures that balance revenue management, resource efficiency, and fiscal sustainability, including:

- *Building Better Water Rates for an Uncertain World:* A handbook with guidance and real-world examples on building an efficiency-oriented rate structure, implementation, public engagement, and financial planning (AWE 2014).
- AWE Sales Forecasting and Rate Model (AWE 2016): "A user-friendly analytical tool that can explicitly model the effects of uniform, block, and seasonal rate structures. Typical water rate models assume that future sales are known with certainty, and do not respond to price, weather, the economy, or supply shortages—that is to say, not the world we live in" (AWE 2016). The model addresses this shortcoming and makes it possible to analyze variable consumption connected to weather extremes, demand response using price elasticity data, drought pricing, and revenue risk simulation over five years (AWE 2016). These resources can be found at www.FinancingSustainableWater.org, a website that brings together research and resources for rate setting. It also provides specific guidance and information for finance managers, conservation coordinators, elected officials, and other stakeholders.

MAINTAINING AN ADEQUATE BUDGET

Water conservation programs are often viewed as a cost center. Sometimes the benefits produced by the program are misunderstood. A good water conservation plan should include a five-year budget, justified by a cost-effectiveness analysis. There are steps the conservation manager should take to ensure adequate funding.

Marketing Program Success and Needs to Finance Managers

Because water conservation is budgeted annually or every two years, along with other utility departments, and sometimes because of the lower-revenue issues, it can be difficult to obtain an adequate budget to carry out planned programs. The process can consume a significant amount of the water conservation manager's time. It is especially difficult to increase budgets dramatically year after year or to add staff. Therefore, when the plan is adopted, the required staffing and budget must be highlighted to decision makers, rather than hidden.

The water conservation manager should try to market the program to senior management, because they may not retain prior agreements on budget and staffing. One way to do this is to publicize the program's successes. This can be done using a number of techniques and venues, including the following:

- Involve political officials in program kickoffs;
- Conduct water-savings analyses and public opinion polls to document success;
- Quickly terminate program elements that are not meeting expectations; Do not have too many programs going at once; successfully complete programs before starting new ones;
- Brief senior management on program progress at least monthly, include how the program is benefiting the utility, such as saving money on water costs;
- Routinely report (that is, every six months) to the board of directors about program progress and successes;
- Stretch the utility's dollar by leveraging partnerships into financial or in-kind contributions to the program costs; and
- Garner public support for the program (as presented in chapter 6).

KEEPING THE PROGRAM ON TRACK TO MEET GOALS

Water utilities should periodically evaluate water savings from selected completed programs to evaluate the long-term effects of program implementation. New information should be used routinely to recalculate the cost of water saved. The cost of water saved can be expressed as dollars per 1,000 gallons saved. In most cases, the programs should continue until the costs, implementation rates, water savings, or other factors change and cause the cost of water saved to rise above some predetermined threshold. This threshold could be the cost of meeting objectives or obtaining a new water supply by some other means. When a program is no longer cost-effective, it should be terminated and the resources should be placed elsewhere.

In addition, new water-saving technology is continually being developed and used by innovative water utilities. Results from the use of these new technologies are often reported at conferences such as AWWA's Sustainable Water Management and at WaterSmart Innovations. Utilities can consider replacing some of their current or no longer cost-effective program elements with new technologies that better fit their situation.

Water utilities should always be flexible when tailoring and revising programs to fit current local conditions. This is usually done as part of the utility's annual budgeting process. Periodically, such as every 5 to 10 years, a major reevaluation of goals and direction is warranted and a new water efficiency plan is developed using the methods described in this manual.

REFERENCES

Alliance for Water Efficiency (AWE). 2014. "*Building Better Water Rates for an Uncertain World.*" http://www.financingsustainablewater.org/tools/building-better-water-rates-uncertain-world, accessed Oct. 18, 2017.

Alliance for Water Efficiency (AWE). 2016. "AWE Sales Forecasting and Rate Model." http://www.financingsustainablewater.org/tools/awe-sales-forecasting-and-rate-model (accessed Feb. 26, 2017).

American Water Works Association. 1993. *Evaluating Urban Water Conservation Programs: A Procedures Manual.* Denver, CO: AWWA.

AWWA. 2004. Manual M54, *Developing Rates for Small Systems.* Denver, CO: AWWA.

AWWA. 2012. Manual M1, *Principles of Water Rates, Fees, and Charges*. Denver, CO: AWWA.

City of Santa Cruz. 2017. Water Master Plan. January.

Raftelis Financial Consulting/American Water Works Association. 2016. Water and Wastewater Rate Survey. Denver, CO: AWWA and RFC.

Saving Water Partnership. 2013. http://www.savingwater.org/cs/groups/public/@spu/@swp/documents/webcontent/01_031210.pdf

Appendix A

Case Studies

Geographic Location	Topic	Supports Chapter	Location
Phoenix, AZ	State of Arizona Municipal Water Use Requirements	Chapter 2	Case Study #1
Tampa Bay, FL	Long-term Supply / Demand Management Plan	Chapter 3	Case Study #2
Philadelphia, PA	Water Loss and Water Loss Control in Philadelphia, Pennsylvania	Chapter 3	Case Study #3
Abbotsford, BC	Using Advanced Metering Infrastructure to Enhance Conservation Efforts in Abbotsford, B.C., Canada	Chapter 3	Case Study #4
Seattle, WA	Water Smart Technology Program: Industrial, Commercial and Institutional Program	Chapter 3	Case Study #5
Fort Collins, CO	Assessments Lead to Customer Action: Sprinkler System Audit Program - Landscape	Chapter 3	Case Study #6
Cobb County, GA	Cobb County Water System's Give them an Inch...Grow a Yard Program	Chapter 3	Case Study #7
New York, NY	New York City Challenges Hotels to Save Water	Chapter 3	Case Study #8
Guelph, Ontario	Graywater Reuse and Rainwater Harvesting Rebate Programs	Chapter 3	Case Study #9
Austin, TX	Mature Conservation Program Shifts Focus From Rebates to Codes due to Saturation	Chapter 4	Case Study #10
Waukesha, WI	Comprehensive Municipal Water Conservation Plan for Mid-Size Suburban Community	Chapter 5	Case Study #11
Sacramento, CA	Engaging Stakeholders in Water Conservation Planning	Chapter 6	Case Study #12
Town of Cary, NC	Block Leader Program - Environmental Program for Public Education of Water Customers	Chapter 6	Case Study #13

CASE STUDY #1: PHOENIX, ARIZONA – STATE OF ARIZONA MUNICIPAL WATER USE REQUIREMENTS

In the 1970s, the state of Arizona was facing a water crisis. Surface water supplies had been fully utilized and groundwater levels were falling rapidly because of agricultural and municipal pumping far in excess of the natural recharge rate. Arizona had rights to the Colorado River, shared with California, Utah, Nevada, and Colorado, but the water was hundreds of miles away from the areas in need. The federal government offered to build the canals and pumping stations to bring the water to Phoenix and Tucson, but only if Arizona would agree to reduce groundwater pumping to the safe-yield level by 2025. A state law was passed that created a new state agency, the Arizona Department of Water Resources (ADWR), which devised a plan to reduce groundwater pumping in Arizona's most populous areas, identified and designated as active management areas (AMAs). ADWR planned to reduce groundwater extraction in the following five stages, defined as management periods: 1980–1990, 1990–2000, 2000–2010, 2010–2020, and 2020–2025. As a result, water use per capita has dropped significantly since 2002 and groundwater pumping is declining as use of renewable water supplies increases. ADWR is currently in the adopted fourth management plan, which will be in effect until 2020 or until the first effective date of the fifth and final management plan.

Beginning in 2006, every community water system (~800 total) in the state must prepare, submit, and implement a water conservation plan as part of their system water plan, which also includes drought preparedness and water supply plans. The system water plan is due every five years. (Large municipal water providers located within the state's active management areas are exempt from this requirement because their conservation requirements are more stringent than those for small municipal providers and drinking water systems located outside of the AMAs. A large municipal water provider is a city, town, private water company, or irrigation district that serves more than 250 acre-feet of water per year.) The Groundwater Code requires all water providers in AMAs to submit annual water use reports. Community water systems outside the AMAs have been required to submit annual water use reports since 2006.

During the fourth management period, ADWR regulates large water providers in active management areas in one of two ways – total gallons per capita per day or non–per capita conservation.

Total gallons per capita per day (total GPCD): A large municipal provider regulated under the total GPCD program must limit the annual gallons per capita per day water usage within its service area to the amount allowed under its total GPCD requirement. ADWR is considering setting the total GPCD requirement for a large municipal provider for the fourth management period as the provider's median total GPCD for the period 2000–2009 minus two standard deviations. However, if the median total GPCD minus two standard deviations is less than the provider's minimum total GPCD requirement, the provider's total GPCD requirement shall be the minimum total GPCD requirement. Minimum total GPCD is based on trends in landscape water use in new developments and WaterSense fixture standards. ADWR is charged with setting a provider's fourth management period GPCD requirement higher than their current requirement under the third management plan.

For each year in which the provider is regulated under the total GPCD program, the actual amount of water withdrawn, diverted, or received by the provider for non-irrigation use will be compared to the amount allowed by its total GPCD requirement to determine compliance during that year. Reclaimed water used directly from a treatment plant or stored underground and recovered within the area of impact of storage is not counted when determining a provider's compliance with its total GPCD requirement.

Also, an infrequent legal type of surface water related to the operation of Roosevelt Dam is not counted.

Total GPCD program compliance is determined pursuant to a flexibility account, which allows providers to use more water than their total GPCD requirement in some years, subject to a maximum negative account balance. Providers who do not meet their targets are subject to fines and sanctions.

Non–per capita conservation program (NPCCP): The non–per capita conservation program is mandatory for all large municipal water providers in active management areas that do not have a designation of assured water supply; it is optional for providers with a designation of assured water supply. The NPCCP requires participating providers to implement best management practices (BMPs) that result in water use efficiency in their service areas. A water provider regulated under the program must implement a required basic public information program and choose one or more additional BMPs based on its size as defined by its combined total of residential and nonresidential water service connections:

- Tier 1 – up to 5,000 service area connections: one additional BMP,
- Tier 2 – 5,001–30,000 service area connections: five additional BMPs, or
- Tier 3 – more than 30,000 service area connections: 10 additional BMPs

The 53 BMPs are divided into the following seven categories: public awareness; education and training; outreach services; physical system evaluation and improvements; ordinances, conditions of service, tariffs; rebates/incentives; and research/innovation.

Providers regulated under the NPCCP are also required to submit a conservation efforts report along with their annual water withdrawal and use report, submit a current rate structure, and submit a new provider profile if their tier changes.

History of Municipal Regulatory Programs

Originally, for the first management period, there was only one program, the total gallons per capita per day program. Large municipal water providers were assigned a total system GPCD target based on their 1980 water use, future demand projections, and the potential savings from water conservation measures commonly used in the western United States. The use reduction targets were modest and ranged from 6 to 11 percent over a 10-year period. Generally, the higher the current water use, the higher the target. For the second management period (1990–2000), each provider was assigned a GPCD target based on an extensive analysis of the water use in each provider's area and the conservation potential. During the third management period (2000–2010), GPCD was calculated using a component method that included single-family, multifamily, nonresidential, and lost and unaccounted-for water. The method under consideration for calculating the GPCD requirement during the fourth management period is described above.

Factors such as rapid commercial and industrial development and changing population characteristics affected the ability of some providers to meet their GPCD targets. During the second management period (1990–2000), the alternative conservation program and original non–per capita conservation programs were created. The alternative conservation program was designed to provide flexibility by establishing a limit on groundwater use, a GPCD target for residential water use only, and requirements addressing specific nonresidential uses. Only one provider was regulated under this program, and regulation was discontinued for the fourth management period. The original non–per capita conservation program required providers to eliminate groundwater pumping (unless replaced by groundwater recharge) and to implement reasonable conservation measures (RCMs). Providers regulated under this program agreed to water supply requirements,

Table A-1 History of Regulatory Programs for Large Municipal Providers in Active Management Areas

Management Period	Regulatory Programs Available to Water Utilities
First management period	Total gallons per capita per day (GPCD)
Second management period	Total GPCD
	Alternative conservation program
	Non–per capita conservation program
Third management period	Total GPCD (component method)
	Alternative conservation program
	Non–per capita conservation program
	Modified non–per capita conservation program
Fourth management period	Total GPCD (proposed median method)
	Non–per capita conservation program (same as modified non–per capita conservation above)
Fifth management period	To be determined

implementation of the RCMs in specified ways, an implementation schedule, and detailed annual progress reporting. This program is not included in the fourth management plan due to regulatory change.

During the third management period, state law relating to municipal conservation programs for large municipal water providers was amended when ADWR conducted an evaluation of the regulatory programs for large municipal water providers. In April 2007, legislation was passed to add a new regulatory program to the third management period for AMAs – the modified non–per capita conservation program. The modification to the third management plan became effective in 2008, and the first year of provider program implementation was 2010. During the fourth management period, this program was renamed non–per capita conservation program (because the original non–per capita conservation program had been discontinued) and is described previously.

CASE STUDY #2: TAMPA BAY, FLORIDA – LONG-TERM SUPPLY/DEMAND MANAGEMENT PLAN

Introduction

Tampa Bay Water currently helps meet the water demands of more than 2.3 million people in the tri-county region. Residential demands accounted for nearly 75 percent of billed water consumption, with the remainder associated with the needs of commercial businesses and industry. The agency has been actively involved in quantifying water demand and potential changes in demand through water use efficiency efforts, mainly through member government implementation, since adoption of its original demand management plan in the mid-1990s. Additionally, the agency developed tools to quantify ongoing member water use efficiency programs that helped to meet original planning goals adopted by the Board of Directors.

In 2013, approximately one-half of the water supplies for Tampa Bay Water member governments were dependent on the timing and quantity of local and regional rainfall. In order to meet reliability goals, it is important to understand how variability and uncertainties would affect the planning and development of water supplies. As Tampa Bay Water's reliance on surface water and other alternative water sources continues to increase, the value of increased water use efficiency in managing future long-term supply needs has become evident. As new supply development costs continue to increase, avoided cost of water supply becomes a more critical element of the water supply planning process.

The demand management plan, developed by Hazen and Sawyer for Tampa Bay Water, investigates the benefits and costs of water demand management as a quantifiable, alternative water supply source and is considered one component of the agency's strategic goals to achieve reliability of its water supply and delivery system to our member governments. Through efficient use of available supplies and use of targeted implementation strategies, water use efficiency can help manage peak-day and average-day water demand in conjunction with reducing long-term future water supply requirements. Cost-effective alternatives to new supply development and other valuable benefits can be realized through demand side management, including optimization of existing facilities, deferred capital investment costs, improved public perception, support of future supply projects, and environmental stewardship and protection.

Profile of Regional Water Demand

A regional baseline water demand profile quantifies and describes the water using and economic characteristics of Tampa Bay Water's member government customers. This includes an assessment of water savings estimates achieved from previously implemented conservation programs and the market for water efficiency technologies. The regional profile includes analyses of water use patterns among the major water using sectors in the Tampa Bay region.

Distribution of Water Use

Regionally, there are three major common sectoral uses of water: single-family residential (SF), multifamily residential (MF), and nonresidential (NR), which includes water used by businesses and institutions. The distribution of regional sectoral demands is illustrated in Figure A-1.

Weather-sensitive and weather-insensitive components of single-family demand were estimated regionally and for each member government over Water Year (WY) 2002–WY 2008. Weather insensitive – generally indoor – use is generally influenced by the number of people residing in a household along with the presence and efficiency levels of various indoor domestic end uses (e.g., toilets, washing machines). Outdoor end uses are weather sensitive and tend to be a highly variable component of total water use. Outdoor uses are influenced both by weather and socioeconomic factors. Figure A-2 illustrates the estimated proportion of weather-sensitive demands in the single-family sector by month through time. Annual average single-family household demand over the period 2002–2008 is 229 gallons per day (gpd) and is estimated to include 52 gpd of weather-sensitive and 177 gpd of weather-insensitive demand.

Evaluation of Achieved Water Savings from Existing Programs

Statistical evaluations were undertaken to measure and verify impacts of existing conservation programs implemented by member governments. The results of these evaluations can be summarized as follows:

- **Member government ultra-low-flow toilet rebate programs:** The data indicate households having received one or more rebates used nearly 12 percent less water on average after the change-out of the toilet. Further analyses indicate homes with only one rebate averaged a 10.8 percent reduction.

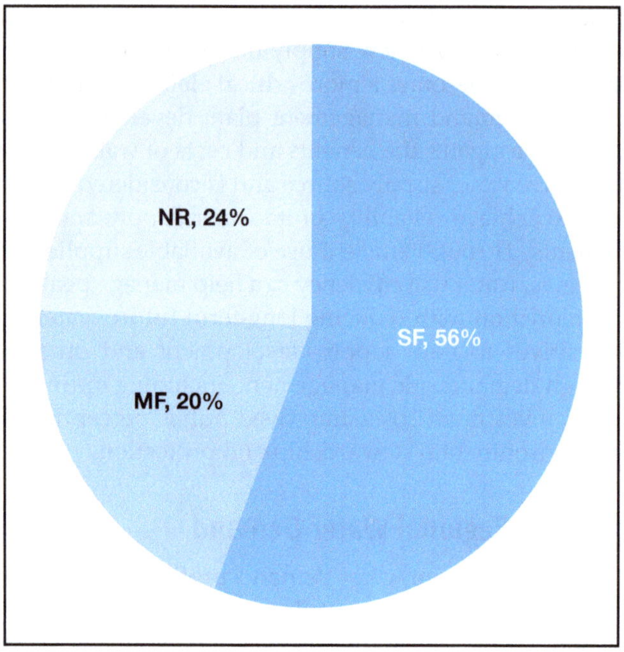

Figure A-1 Distribution of regional sectoral water demands

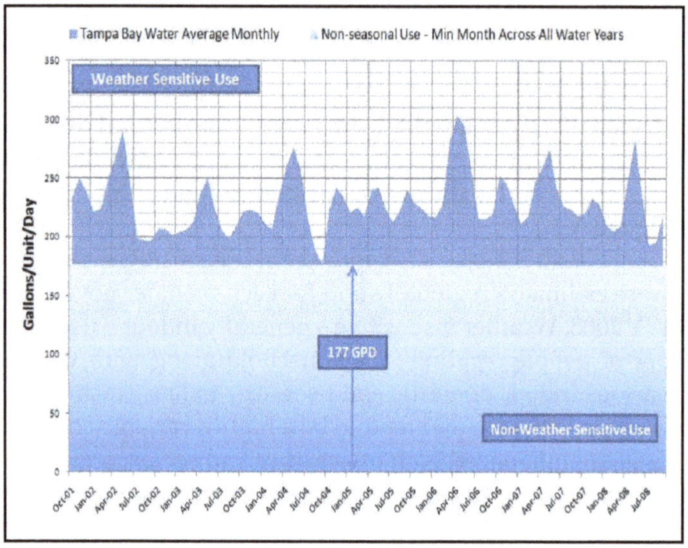

Figure A-2 Regional single-family weather-sensitive and weather-insensitive demands

- **Florida-Friendly Landscaping™:** Homes recognized by the county extension offices as having both water wise landscape design and efficient irrigation technology and practices used about 3 to 5 percent less after one year of participation and from 5 to 9 percent less after two years.
- **Member government irrigation evaluation programs:** Although significant potential may exist, results suggest a diminution of savings over time, with an estimated reduction in water use by about 7 percent after one year of participating and only 3 percent after two years.

Table A-2 Estimated Baseline Single-Family Flow Rates, Gallons per Event (2008)

End Use	Tampa Bay Water	Current Standard	High Efficiency	Estimated % Reduction Under Standard Benchmark	Estimated % Reduction Under High-Efficiency Benchmark
Toilet	2.39	1.6	1.28	–33%	–46%
Shower	2.08	2.5	2.0	20%	–4%
Faucet	1.01	2.2	1.5	117%	48%
Clothes washer[a]	33.49	24.6	15.0	–26%	–55%
Dishwasher[b]	8.90	6.5	6.0	–27%	–33%

[a] Current standard based on 9.5 water factor, 2.7 cubic feet per load, and 0.96 loads per day

[b] Current standard based on federal dishwasher standard effective January 2010

Analysis of Water Technologies and Baseline Efficiency Levels

Through a literature review of available and emerging technologies/programs, a water efficiency program library of technically applicable demand management technologies, programs, and best management practices was developed for potential application in the Tampa Bay region. Examples of residential end-use technologies include toilets, showerheads, faucets, clothes washers, dishwashers, and irrigation. Nonresidential end uses generally include those found in the residential sector, but also consist of technologies that can use substantial quantities of water for cooling, heating, and process water including product development (e.g., food service).

Estimates of water savings potential were based on a changing mix of water using technology, as well as the rate (or intensity) at which water using technology was used. Assessment of technology-based and program-based savings potential required base-year (2008) estimates of distribution of fixture age and efficiency in the region by sector of water use and market penetration of water efficient technologies. These estimates provide a baseline for examining remaining water efficiency potential over the agency's long-term water demand horizon (2035).

Market penetration by passive measures was assumed to be associated with plumbing standards and increased efficiency due to an evolving market (supply and demand) for water efficient products recognized or certified through the US Environmental Protection Agency (USEPA) WaterSense® label, and/or Energy Star® programs.

An end-use assessment of single-family demand in gallons per capita per day along with estimated average end-use flow rates, Table A-2, suggests that the greatest efficiency potential exists in toilet, clothes washer, and dishwasher use.

Evaluation of Water Efficiency Alternatives (Forecast)

Water savings can be realized from either passive or active water use efficiency measures.

- *Passive* water efficiency is achieved through a natural process of replacing old fixtures with new, more efficient fixtures as they wear out or become effectively obsolete, or by installing efficient water using fixtures in new construction due either to codes or market changes.
- *Active* water efficiency measures include programs designed to expedite the replacement process described above. Such programs are often sponsored by water utilities to ensure a target installation rate and associated water savings and can include outdoor efficiency technologies.

Before the potential benefits of active water efficiency alternatives can be assessed, passive savings must be estimated. An assessment of remaining passive efficiency potential was used to identify, develop, screen, and select technically applicable active alternatives.

Passive Fixture and Savings Estimates

The US Energy Policy Act (EPAct), effective in 1994, mandated flow standards for many fixtures (e.g., toilets, faucets, and showerheads, among others). Since then, manufacturers have introduced and marketed fixtures and appliances that far exceed EPAct standards, leading to USEPA WaterSense and Energy Star programming. These programs influence the market by encouraging consumers to purchase high-efficiency (HE) water products. WaterSense-labeled products require independent third-party certification of performance and product durability, ensuring that product use is consistent with labeling over a defined life. As consumers decide to purchase and install HE water products, water consumption efficiency increases.

Accounting for prospective changes in market penetration allows adjustment to the baseline demand forecast reflecting market-based passive demand reductions. Assumptions about efficiency standards, fixture life, and market penetration of HE products were used to estimate fixture distributions and water use for each year in the long-term demand forecast. Passive savings were estimated for residential toilets, washing machines, and dishwashers as well as nonresidential toilets and urinals. By 2035, approximately 26 mgd of water savings potential is estimated and attributable to passive efficiency.

Screening and Selection of Active Efficiency Technologies/Programs

Remaining market potential for water efficient technology (beyond what is likely accounted for by passive measures) was determined through the 2035 demand forecast planning horizon by screening the applicability of several active (utility-sponsored) programs. The screening process included 24 programs/technologies, either applied through existing programs (regionally and nationally) or developed based on specific application of technologies in specific sectors or water end uses. Regional and national literature and other secondary sources, along with information gleaned from survey and analysis of regional water use characteristics, supported the screening process.

The 10 programs meeting screening criteria and selected for inclusion in the demand management plan portfolio are shown in Table A-3. As identified in Table A-3, the most cost-effective program is cooling tower retrofits at an average cost of $0.07/1,000 gallons. The least cost-effective program is the conveyor dishwasher incentive program at an average cost of $0.42/1,000 gallons.

Development of Alternative "With Conservation" Demand Forecasts

Estimated impacts of passive water savings and potential active demand management alternatives on the region's long-term demands were evaluated over the planning horizon. Figure A-3 illustrates the magnitude of estimated water demand reductions from both passive and active savings relative to the 75th percentile baseline demand forecast and current sustainable system capacity. As shown in Table A-4, by 2035 a total of approximately 35 mgd of water use reduction and savings potential was identified. Of this total, 26 mgd of water use reduction is associated with the impact of passive changes, while the estimated additional savings from active efficiency is 9 mgd.

Table A-3 Programs Meeting Screening Criteria

Activity Name	Class	Utility Costs ($/unit)	Savings, Useful Life (yr)	Unit Savings (gal/yr)	Gallons Saved Over Useful Life	$/1,000 gal	Benefit–Cost Ratio
Cooling tower	NR	$1,000	10	1,386,530	13,865,300	$0.07	8.25
PRSV	NR	$30	10	37,426	374,260	$0.08	6.03
HEU (1/2 gallon)	NR	$125	30	18,928	567,853	$0.22	1.26
ULFT (valve-type)	NR	$125	30	17,970	539,100	$0.23	1.30
Alternative irrigation source	SF	$750	25	94,034	2,350,850	$0.32	1.19
HET (tank-type)	NR	$125	30	12,843	385,290	$0.32	0.90
Residential HET	SF	$100	25	11,542	288,550	$0.35	1.11
ET/SMS irrigation controller	SF	$200	10	56,645	566,450	$0.35	1.84
Residential HET	MF	$75	25	8,111	202,775	$0.37	1.03
Conveyor dishwasher	NR	$500	20	59,951	1,199,020	$0.42	1.09

Note: PRSV = pre-rinse spray valves; HEU = high-efficiency urinal; ULFT = ultra-low flush toilet; HET = high-efficiency toilet; ET/SMS = evapotranspiration/ soil moisture sensor; NR = nonresidential; SF = single-family; MF = multifamily

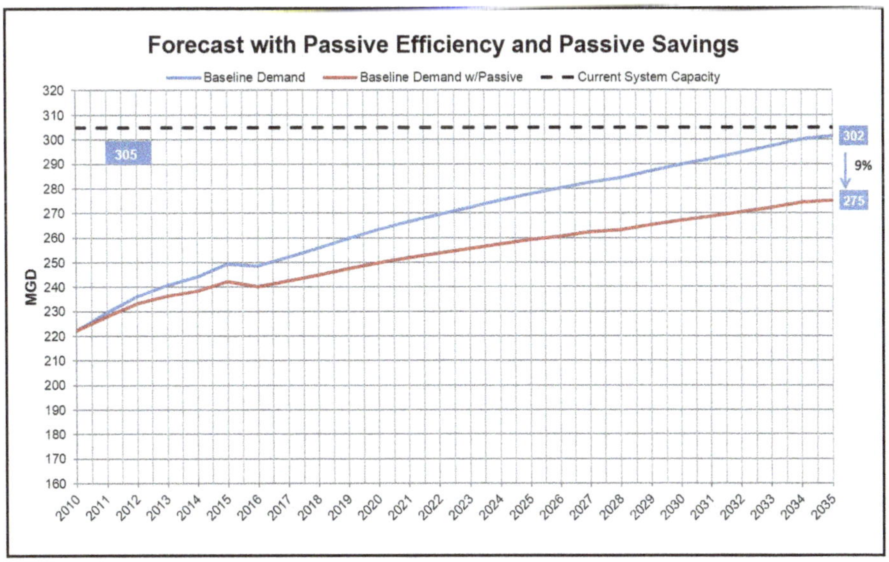

Figure A-3 Baseline demand forecast with passive and active savings

Table A-4 Projected Water Savings from Passive and Active Water Conservation

Forecast Scenario	Projected Water Savings (mgd) / Percent Reduction					
(75th percentile)	2010	2015	2020	2025	2030	2035
Passive savings	0/0	7/2.6	13/4.9	18/6.4	22/7.6	26/8.5
Active savings	0/0	0.3/0.1	4/1.3	7/2.6	10/3.5	12/4.1
Passive and active savings	0/0	7/2.8	16/6.2	25/9.0	32/11	38/12.5

Table A-5 Projected Water Savings from Passive and Active Water Conservation

	Present-Value Cost ($M)	Present-Value Benefit ($M)	Net Present Value ($M)	Benefit–Cost Ratio
Life of savings to 2065	$32.5	$60.6	$28.1	1.85
Life of savings to 2035	$32.5	$42.0	$9.5	1.29

Economic Analysis of Alternative Demand Management Strategies

Quantification of supply-side benefits is based on the accrual of avoided costs. Significant benefits can demonstrate how proposed efficiency measures can defer source development and save money. Thus avoided costs (or benefits) from water use efficiency can be categorized as resulting from

- Capital deferral
- Capital elimination
- Reduction in variable cost

Tampa Bay Water determined savings and costs over a 60-year planning horizon (2010–2069), allowing savings rates in this analysis to mature over the life of the technology installed. Net avoided costs of viable demand management alternatives were evaluated over two separate time frames: the total life of all savings and through the 2035 forecast horizon. See Table A-5.

CASE STUDY #3: PHILADELPHIA, PENNSYLVANIA – WATER LOSS AND WATER LOSS CONTROL IN PHILADELPHIA, PENNSYLVANIA

The Philadelphia Water Department (PWD) operates one of the largest water supply systems in the United States with more than 3,000 miles of water distribution pipelines and nearly half a million customer accounts. Philadelphia is blessed with relatively abundant water resources, but it has one of the oldest water distribution systems in the country and a population that has declined by one-third since 1950. While concerns for long-term source water availability are not paramount for PWD, water efficiency is important in order to contain operating costs and generate sufficient revenue in a service area with affordability concerns for a notable segment of the customer population. Hence, PWD has focused heavily on the supply-side practice of *water loss control,* which seeks to optimize the distribution of drinking water and maximize the billing and revenue potential for the utility.

Water loss control keys on both accountability of the water supply (which is achieved by conducting an annual system water audit) and efficiency (by controlling to economic levels both apparent [non-physical] losses and real [physical] losses). Since the early 1990s, PWD has dedicated staff and funding to develop and implement sound water loss control practices, and has become successful in slowly improving the water efficiency of the large urban water supply system that it operates. In addition to its internal efforts, PWD is heavily involved in drinking water industry activities. It has participated in the development of guidance publications, software, and research to further the advancement of water loss control methods used by water utilities worldwide.

The Water Audit: The Foundation of Effective Water Supply Efficiency

The water audit is an accounting approach used to track the volumes of treated drinking water managed by a water utility from the point of entry to the water distribution system, during its conveyance across the system, and through recording and billing at the customer endpoint. Compiling the water audit is the best means to establish accountability of supply in utility operations, and it is recommended that all water utilities assemble a representative water audit on an annual basis as a standard business practice.

Staff from PWD and the city's Water Revenue Bureau formed the city's Water Accountability Committee in the early 1990s, and the committee issued its first annual water audit for the 1998 fiscal year (ending June 30, 1998). In 2000, Philadelphia became the first water utility in the United States to compile an annual water audit using the method published by the International Water Association (Alegre et al. 2000). In 2006, Philadelphia became one of the first users of the American Water Works Association (AWWA) Free Water Audit Software© (AWWA 2014a) and has volunteered its water audit data for review and validation by AWWA's Water Loss Control Committee since the launch of its Water Audit Data Initiative (AWWA 2014b) in 2011. The AWWA Free Water Audit Software was created and is maintained by the AWWA Water Loss Control Committee and is available for free download from the AWWA website (AWWA 2014a, 2014b). The site also includes considerable additional information on water loss control. PWD recognizes the value of having an objective and representative water audit, and has been a consistent early adopter of new methods and tools for improved water auditing.

The AWWA water audit method is recognized as the best practice water audit method endorsed by AWWA and is detailed in AWWA manual M36, *Water Audits and Loss Control Programs* (AWWA 2015). Readers should refer to this publication and the AWWA Free Water Audit Software for specific guidance on the use of this method. The crux of this method is that it keys on quantifying the volumes of water consumed and water lost, and the cost impacts of the losses, both on utility operations and as revenue shortfall. The method includes an array of objective performance indicators with emphasis on effective benchmarking and performance tracking indicators. Less focus is directed to percentage indicators, which are known to be imprecise in reflecting operational efficiency.

For water utilities that have not traditionally compiled an annual water audit, the development of a representative and reliable water audit usually requires three to five years of increasingly refined procedures for data collection and validation, leading to audit results that are reflective of utility operating performance. The AWWA Free Water Audit Software includes a data grading feature that allows the auditor to grade or rate the validity of the data inputs, from a low grade for less reliable data to higher grades for accurate, well validated data. The component gradings are used to calculate a composite data validity score that reflects the overall integrity of the data in the water audit. The data validity score is based on a scale of 1–100, and utilities can use this score as a measure of the trustworthiness of their data. The software includes a data grading matrix that offers guidance on grading assignments based on the process attributes of the utility. By improving their water supply processes, water utilities can produce more reliable data and a higher data validity score. But, just as important, these improved processes often lead to expedient water efficiency improvements. Thus the water audit is not just an exercise in data collection, but can be an evolution in progress that creates improvements in the water utility's water supply efficiency year by year.

PWD's annual water audit has evolved in just such a manner over a period of more than 15 years. A summary of PWD water audit statistics for several recent years is given in Table A-6. It is important to recognize that PWD sustained average daily non-revenue water (NRW) levels of 120–140 mgd each year as recently as the mid-1990s. Table A-6 shows NRW levels ranging from 73 mgd to 91 mgd in the six-year period listed; thus PWD operates with NRW levels approximately 40–60 mgd less than it did 20 years ago.

In terms of physical water savings, PWD's water withdrawals from its two river sources peaked at an annual daily average of 393 mgd in 1981 and were measured at 259.8 mgd in 2013. This is a difference of 135 mgd, or 49.3 billion gallons of water saved annually. This reduction in source water withdrawal is attributed not to a single reason but is a combination of improved distribution system leakage control, a smaller customer population, gradual installation of low-flow fixtures in customers' homes, and a move from quarterly to monthly billing with reliable customer meter readings collected by an efficient automatic meter reading (AMR) system. The improved accuracy of meter reading and billing data has allowed customers to better manage their water usage. The per capita use/loss of *all* water on a systemwide basis (this includes both customer consumption and losses) has declined from 196 gallons per capita per day (gpcd) in 1980 to 148 gpcd in 2013. These per capita figures are derived by dividing the total source water withdrawal volume by the total service area population.

Overall, PWD has achieved notable success in reducing water withdrawals and water loss over the long term. Still, with current NRW between 70 and 90 mgd, PWD continues to operate with losses above economic levels, and water loss control remains a major focus for the utility. The importance of the need to conduct a water audit on an annual basis is proven by cost data trend shown in Table A-6. Apparent losses are typically valued at the composite unit rate that a water utility charges its customers for water service.

Thus, each time the water utility implements a water rate increase, the unit cost of its apparent losses also increases. Similarly, the unit cost of real (leakage) losses is often valued at the variable production cost (although real losses could be valued at the customer retail rate in areas where available water resources are, or will be, in deficit.) Using the variable production costs as the basis for real loss valuation, whenever the costs increase for water treatment, electric power for pumping, and related production, then the cost of leakage increases commensurately. Since water utilities rarely enact a water rate decrease, the unit costs for apparent losses only tend to increase over time.

Similarly, the variable costs of water production in a utility only increase over time. While a water utility may have greater or lesser success in reducing its losses from year to year, the unit costs of the losses are always on the rise. Hence the water utility is behooved to assemble the water audit on an annual basis – just like its financial audits – and to reassess both the loss volumes and the loss cost impacts, both of which serve as the basis for the loss control program strategy.

Controlling Apparent Losses: Establishing a Revenue Protection Program

Apparent losses include three primary components: unauthorized consumption, customer metering inaccuracies, and systematic data handling error in customer billing systems. Apparent losses comprise water that has reached a customer but has not been (fully) measured or billed. Apparent losses collectively understate the consumption by the customer population and result in utility underbilling and loss of revenue collection potential for a portion of the population. PWD encounters notable volumes of apparent losses in all three components.

As part of the water audit process, it is very important for the water utility to first come to understand the inner workings of the customer billing system, particularly as it relates to customer consumption volumes measured by water meters. Many billing

Table A-6 Annual Water Audit Summary Data for the Philadelphia Water Department for Fiscal Years 2008 to 2013

	Fiscal Year					
	2013	2012	2011	2010	2009	2008
Water Supplied						
Water Supplied from Rivers, mgd (minus treatment plant usage)	259.8	257.9	271.7	266.9	268.0	273.6
Water System Input Volume, mgd	238.4	237.1	250	244.4	245.8	250.7
Service Area Population	1,758,507	1,747,371	1,736,906	1,736,906	1,660,900	1,660,534
Per capita water supplied, gpcd	147.7	147.6	156.4	153.7	161.4	154.8
Authorized Consumption						
Billed Consumption, mgd	152.1	152.6	158.4	165.8	172.4	175.8
Unbilled Unmetered (firefighting, etc.), mgd	5.4	3.3	2.1	2.0	2.1	2.1
Unbilled Unmetered water cost	$1,398,440	$1,051,979	$866,917	$779,394	$713,802	$717,246
Authorized Consumption, mgd	157.4	155.9	160.5	167.8	174.5	177.9
Per capita water billed, gpcd	86.5	87.3	91.2	95.5	103.8	105.9
Water Losses						
Apparent Losses (billing data error, unauthorized consumption, customer metering inaccuracies), mgd	20.5	21.4	23.1	17.0	15.0	19.0
Apparent Loss unit retail cost/mgd	$7,313	$6,983	$6,503	$6,071	$5,429	$4,971
Apparent Loss Cost	$43,097,617	$43,076,304	$43,544,119	$30,034,457	$22,255,068	$27,328,369
Real Losses (Leakage), mgd	60.4	60.0	66.3	59.6	56.2	53.8
Variable Production Cost, per mgd	$346.46	$345.86	$272.56	$235.00	$229.41	$215.50
Real Loss (leakage) Cost	$8,153,737	$8,099,949	$7,357,751	$5,867,776	$4,843,686	$4,960,676
Non-Revenue Water Volume, mgd = Unbilled, Unmetered Consumption + Apparent Loss + Real Loss	86.3	84.6	91.6	78.6	73.4	74.9
Non-revenue Water Cost Total	$52,649,794	$51,390,813	$50,901,870	$36,681,627	$27,098,754	$33,006,291
Key Performance Indicators						
Apparent Losses per connection, gal/service connection/day	38.9	39.8	44.4	30.8	25.8	35.2
Real Losses (leakage) per connection, gal/service connection/day	111.6	11.9	126.2	107.7	96.6	96.7
Infrastructure Leakage Index (ILI)	8.8	10.2	11.5	9.9	8.9	9.0
Non-revenue Water by Volume, %	38.5%	38.4%	39.7%	34.9%	32.3%	32.4%
Non-revenue Water by Cost, %	21.1%	22.2%	23.2%	16.3%	12.5%	15.2%

systems employ multiple adjustment routines or other financial accounting manipulations of data in the billing system. If adequate controls do not exist for these actions, lapses can occur and result in understated billings to customers. Such instances represent systematic data handling error, and this is the largest component of apparent loss for PWD. PWD continues to meet regularly with staff from Philadelphia's Water Revenue Bureau, which conducts water billing in the city. A number of procedural improvements have been incorporated to ensure that all customers are properly billed. However, in a large system of nearly half a million accounts, there will always be some billing error.

Unauthorized consumption is also a significant issue for PWD. Occurrences include customer meter tampering, AMR equipment tampering, illegal restoration of services shut off for nonpayment, illegal connections, and illegal fire hydrant openings. Water utilities can address the issue of unauthorized consumption by assessing policy and regulations, implementing means to detect such occurrences, enacting penalties for infractions, and devising means to thwart continued incidents. PWD has achieved success in employing center compression locks on its fire hydrants to deter unauthorized hydrant openings. Its AMR system has tamper detection capabilities that flag illicit actions.

PWD reviews regular occurrences of zero-consumption accounts in the billing process. Vacant buildings account for some zero-consumption bills, but in roughly one-third of these instances, accounts register zero consumption for several months because the customer meters and/or metering reading equipment have been tampered with or damaged. PWD is reviewing its policies to determine whether more targeted or stern penalties should be enacted for repeat offenders. The extent of unauthorized consumption varies widely among utilities, but large cities tend to encounter this problem to a greater extent than smaller systems. However, all water utilities are susceptible to some amount of unauthorized consumption, and this occurrence should be considered as part of the water audit process.

As part of its AMR system installation between 1997 and 1999, PWD replaced all of its residential customer meters. Meter accuracy testing has confirmed that the expected life of most of these meters will exceed 20 years; hence PWD is highly confident in the accuracy of this meter population for the near-term future. Conversely, the accuracy of the population of large commercial and industrial meters is suspect. PWD, like most water utilities, uses mostly turbine and compound meters in these applications. PWD has conducted extensive data-logging on several hundred such accounts between 2006 and 2013 and has found that many meters are the wrong type for the actual flow profile and many are oversized, or both. Most often, such inappropriate installations result in notable volumes of water occurring at flow rates that are too low for the meter to register, which results in underbilling of these large customers.

PWD has begun to investigate and pilot certain meter types as alternatives to the traditional turbine and compound meters, by conducting trials of single jet meters and (soon) magnetic flowmeters in some of these applications. Since commercial and industrial customers typically consume a significant portion of the total customer consumption, it is important for a water utility to assess the reliability of the large customer meters. PWD encounters significant missed revenue potential due to its apparent losses and is continuing to direct efforts to better define and control these losses.

Controlling Real Losses: Leakage and Pressure Management

For more than 30 years PWD has operated a proactive leak detection program, with both day and night crews, to continuously survey the water distribution system searching for hidden, unreported leaks. Additionally, it has been keen in embracing innovative technology that moves away from historical, reactive approaches to leaks and toward proactive methods. PWD's standard leak detection efforts include crews using electronic sounding equipment to detect the presence of leaks and electronic correlating equipment to pinpoint the leak sources. Crews survey the distribution system on a schedule that covers the entire system on average roughly once every three years. However, certain high-leakage areas are surveyed annually, and certain low-leakage areas are surveyed perhaps once every five years. This type of acoustic leak detection program is the most common in the North America water industry. It is highly recommended that all water utilities conduct acoustic leak detection surveys as frequently as is appropriate for their system size and occurrence of leakage.

While acoustic leak detection is the foundation of leakage management, manned leak detection surveys also have limitations, and PWD has sought new technologies that address these shortcomings. Traditional leak detection surveys are less effective on large-diameter transmission pipelines than on smaller water distribution pipelines. In 2007 PWD contracted for use of inline transmission pipeline leak detection services in order to better assess leakage occurring on its large-diameter pipelines. Inline leak detection is a newer technology that sends an acoustic sensor into an active pipeline to listen for leaks. While the technology has certain logistical limitations, it nonetheless has been highly accurate in pinpointing leaks on a number of large-diameter pipelines in Philadelphia.

Table A-7 Program Summary for Inline Leak Detection for Large-Diameter Transmission Pipelines in Philadelphia from 2007 to 2014

Fiscal Year	Period	Total Mileage for Program	Number of Confirmed Leaks Identified	Number of Leaks Per Mile Surveyed	Cost
2014	5/6–5/13/2014	6.54	5	0.8	$135,000
	11/14/2013	0.19	0	0.0	$15,000
2013	6/4/2013–6/8/2013	2.36	1	0.4	$75,000
	10/23/2012–10/27/2012	3.77	5	1.3	$75,000
2012	5/8/2012–5/18/2012	6.34	12	1.9	$150,000
2011	10/4/2010–10/8/2010	2.96	6	2.0	$60,000
2010	5/24/2010–5/28/2010	2.76	10	3.6	$60,000
	10/19/2009–10/23/2009	1.96	3	1.5	$60,000
2009	6/1/2009–6/5/2007	2.78	3	1.1	$48,000
	11/3/2008–11/14/2008	5.61	14	2.5	$108,000
2008	4/14/2008–5/1/2008	7.15	8	1.1	$101,738
	10/15/2007–10/19/2007	3.1	15	4.8	$60,000
2007	6/4/2007–6/8/2007	3.53	3	0.8	$50,000

Table A-7 provides a summary of PWD's program for inline transmission pipeline leak detection. PWD has benefited from the use of this technology not just in terms of preventing lost water, but also in lessening the likelihood of catastrophic failure of these pipelines. As shown in Table A-7, the leak detection service provider has scanned more than 49 miles of large-diameter pipelines over an eight-year period, with 85 leaks detected and repaired. This technology has identified an average of two leaks per mile on Philadelphia's large-diameter pipelines. Many of the leaks detected are small, low-volume leaks. Thus, PWD has been able to repair these leaks before they could become large leaks or disruptive ruptures. Use of this service addressed a previously unanswered need for PWD: effective leak detection for large-diameter pipelines. The average cost per mile for the eight year PWD program was $20,341 and the average cost per leak was $11,738.

Other successful technologies for large pipeline leak detection are becoming available. It is recommended that water utilities stay abreast of developments in this area and consider undertaking such work on their most critical pipelines.

PWD has also piloted successful international leakage management technology by its participation in the research project described in Leakage Management Technologies (AwwaRF 2007) and administered by the Water Research Foundation (formerly AwwaRF). As part of this project, PWD designed and installed a permanent district metered area (DMA), which is a discrete zone in the water distribution system isolated to one or two supply mains. This allows for close monitoring of water supply flows to distinguish volumes of customer consumption from leakage.

As part of the project, PWD designed and installed a permanent DMA in an older area of its water distribution system. A boundary of closed valves was established and two supply pipelines – one routine and one standby – provided water flow to the DMA. The flow is continuously measured, monitored, and archived by PWD. In the eastern United States, customer irrigation systems are not common: During the nighttime hours in residential zones or DMAs, the supply flow drops as legitimate demand diminishes; the bulk of the night flow is leakage. By measuring and analyzing these "nightline flows, PWD is able to discern the composite leakage rate occurring in the DMA. Since the flow is continuously monitored, PWD can see when leakage levels are beginning to rise and can strategically direct crews to conduct acoustic leak detection.

In addition to the enhanced monitoring provided by the DMA, PWD demonstrated advanced pressure management as an approach for both leakage control and long-term sustainability of the water pipeline infrastructure. PWD's DMA is located in an area of its water distribution system that is supplied with relatively high water pressure of over 100 pounds per square inch (psi). Established relationships exist among water pressure level, leakage rates, and incidences of water main rupture. PWD set out to demonstrate that improved management of the excessive pressures in this DMA would provide enhancements in reduced leakage and, perhaps, a reduction in the long-term water main break occurrences in this area.

PWD installed pressure-reducing valves on each of its two sources and implemented a flow-modulated pressure management scheme that lowered the average DMA pressure by about 35 psi (from roughly 110 psi to 75 psi). In this scheme, pressure is paced with flow into the DMA, decreasing at night when customer demand declines, and increasing during daytime hours when customer consumption rises. In this way leakage from pressure-sensitive "background" leakage is starved during the night hours, thus resulting in leakage reduction. PWD has monitored flows in the DMA continuously since its full commissioning in 2007 and has achieved significant leakage reduction by a combination of advanced pressure management, several rounds of acoustic leak detection, and selective water main replacement work. PWD quantified this leakage reduction from a measured volume of 566 gallons per customer service connection per day to 44 gallons/connection/day by 2009.

Leakage is dynamic: new leaks are always forming even as existing ones are repaired. Since 2009 leakage has risen slowly, but PWD is able to monitor problem areas closely and intervene to keep the overall leakage level contained to a reasonable minimum. PWD continues its analysis in this area by monitoring the rate of occurrence of water main breaks. Since water main breaks occur less frequently than leaks, a longer timeline is needed to make a reliable judgment on the benefits of pressure management in reducing main breaks. PWD envisions compiling a ten-year record of breaks before reaching any conclusions on this trend. A detailed description of this project and its findings is available in the literature (Kunkel and Sturm 2011).

The use of DMAs is new to the North American water industry but has become standard practice in a number of countries, including in the United Kingdom. It is an effective technique that moves water utilities from retrospective, reactive responses to leakage to proactive control of leakage. PWD benefited from its research on this technique. District metered areas can have greater or lesser benefits depending on distribution system characteristics, and PWD has begun monitoring a second DMA since the success of its first permanent DMA.

Summary

Accountability and efficiency of water supply operations are necessary if water utilities are to act as good stewards of water resources. Managing customer demand through a well-structured water conservation program is an effective means to optimize water consumption by end users. Additionally, water suppliers should be efficient in their operations. Water loss control is a collection of methods and practices that water utilities can employ to ensure that the physical losses from water leakage are reasonably contained, and that the nonphysical accounting and measurement losses are likewise managed. Employing water loss control requires a shift in mindset by water utility managers: from reactive to proactive monitoring and from outdated to robust practices. The Philadelphia Water Department has been at the forefront of the development, application, and analysis of innovative methods for water loss control for more than two decades.

References

Alegre, H.; W. Hirner; J. Baptista; and R. Parena. 2000. *Performance Indicators for Water Supply Services.* International Water Association (IWA) Manual of Best Practice series. London: IWA Publishing.

American Water Works Association (AWWA). 2014a. "Free Water Audit Software, version 5.0." AWWA Water Loss Control Resource Community. https://www.awwa.org/resources-tools/water-knowledge/water-loss-control.aspx, accessed February 28, 2017.

AWWA. 2014b. "Validated Water Audit Data." AWWA Water Loss Control Resource Community. www.awwa.org/resources-tools/water-knowledge/water-loss-control.aspx, accessed February 28, 2017.

AWWA. 2015. *Water Audits and Loss Control Programs,* 4th ed. Manual M36. Denver, CO: AWWA.

AWWA Research Foundation (AwwaRF). 2007. *Leakage Management Technologies.* Denver, CO: AwwaRF.

Kunkel, G.; and R. Sturm. 2011. "Piloting Proactive, Advanced Leakage Management Technologies." *Journal AWWA* 103(2):62–75.

CASE STUDY #4: ABBOTSFORD, BRITISH COLUMBIA, CANADA – USING ADVANCED METERING INFRASTRUCTURE TO ENHANCE CONSERVATION EFFORTS

Background

Water conservation is becoming extremely important in some areas of Canada. While rainfall is plentiful through the majority of the year in the Fraser Valley, British Columbia, rapid growth in the past 15 years has put a strain on infrastructure in some areas. The Abbotsford Mission Water and Sewer Commission (AMWSC) governs the water system for both Abbotsford and Mission. (The AMWSC is made up of councilors and the mayors from both Abbotsford and Mission.) Abbotsford has a population of approximately 144,000 and Mission 39,000, with a total serviced population of about 160,000, of which Abbotsford uses approximately 75 percent of the share. With demands increasing rapidly from 2001 through 2007, the AMWSC has watched regional peak-day water demands carefully since then.

In 2006 and 2007, water demand during peak day was within 98 percent of the system capacity. While average day demands were easily manageable throughout these years, the peak-day demand was up to 80 percent greater than the average-day demand. Although a peaking factor of 1.8 is common in many areas – and often higher in other municipalities – the issue of rapid development and limited system capacity was cause for concern. Based on demand projections, it was predicted that a new water source would be needed by 2016.

In response, the AMWSC approved a three-pronged approach in 2010 for the future of the water system: identify a new water source, squeeze more capacity out of the current system, and focus heavily on water conservation.

A new source was identified and significant effort was invested to optimize existing infrastructure. As a short- to medium-term solution, the AMWSC developed additional groundwater supply to supplement existing sources. Until the new water source was

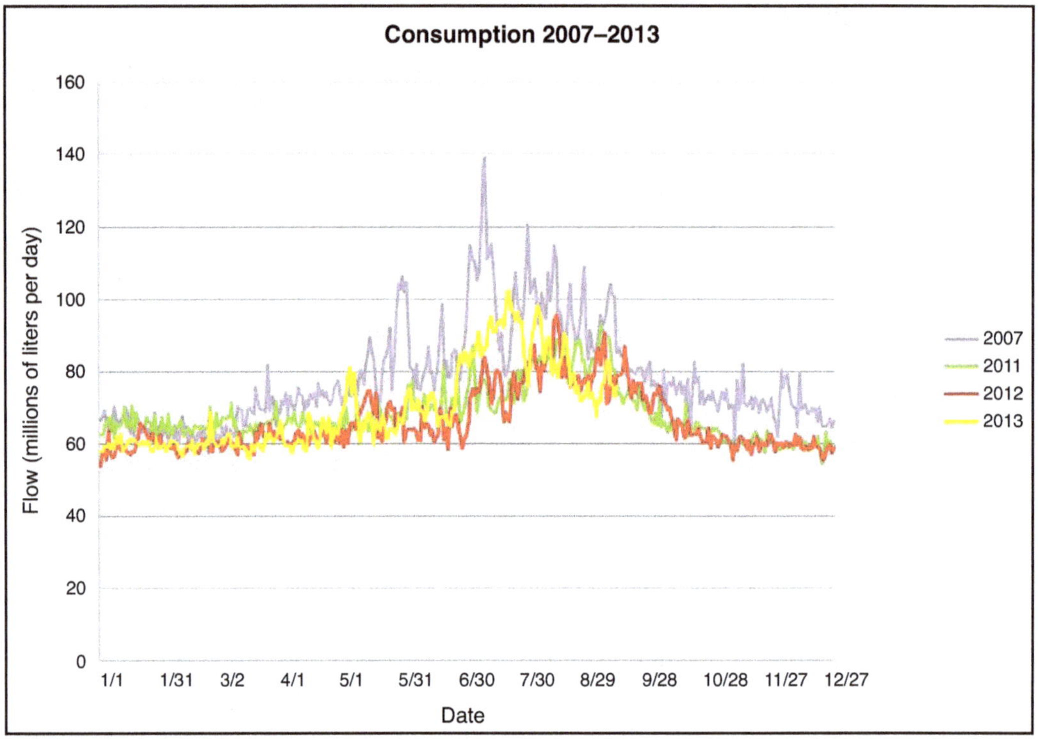

Figure A-4 Water reduction from 2007

online, conservation became extremely important. Using 2007 as a baseline year, a target of 25 percent in reduction of peak demand was set for 2015.

In 2011, the year after the water master plan was approved, the water use reduction goal was exceeded without implementing a full watering ban. The region saw a 33 percent reduction in peak-day demand and a 12 percent reduction in average-day demand, relative to 2007 (Figure A-4). The numbers remained similar in 2012 with a slight increase in 2013.

Discussion

Much of the success came from building a solid foundation for conservation in Abbotsford to support the programs offered. However, it should be noted that 2011 and 2012 had cooler summers. Metering, frequent billing, and conservation-oriented pricing are fundamental tools of water system management and conservation. This fundamental structure helped provide customers with the knowledge and financial incentives to use water efficiently and take advantage of the conservation programs in place.

At the municipal level, Abbotsford has been universally metered since 1995. In 2010 its city council approved a new metering project, and Abbotsford was one of the first cities in Canada to install a two-way-communication, fixed-base-network advanced metering infrastructure (AMI) system (see Figure A-5). This system has provided numerous benefits to the city, including:

- Implementing frequent billing and conservation-oriented pricing
- Enhancing conservation programs
- Evaluating the benefits of conservation
- Analyzing demands for future conservation programs

Figure A-5 Itron automated meter end points used in Abbotsford

Table A-8 Seasonal Increasing Block Rate Structure

Off-Peak Rate – November to April	\$1.13/m³ (\$3.20/hcf)
Proposed Seasonal Tiered Rates – May to October	
Tiers (based on bimonthly consumption)	
Tier 1 0–60 m³ (0–21.2 hcf)	\$1.13/m³ (\$3.20/hcf)
Tier 2 61–90 m³ (21.5–31.8 hcf)	\$1.43/m³ (\$4.05/hcf)
Tier 3 91 m³ + (32.1 hcf +)	\$2.26/m³ (\$6.40/hcf)

Note: m³ = cubic meters; hcf = hundred cubic feet
All \$ values are in Canadian dollars.

Because of the difference in metering infrastructure between Mission and Abbotsford, most of the AMWSC's success in water conservation to date has been in Abbotsford. The majority of customers in Mission are not metered and thus great potential exists to reduce demand. In 2015, Abbotsford's per capita demand was 185 liters per day (53 gallons per day), reduced from 281 liters per day (74 gallons per day) in 2007. Mission's per capita demand in 2015 was estimated at 335 liters per day (88 gallons per day).

Frequent Billing and Conservation-Oriented Pricing

Shortly after the installation of the AMI system in 2011, billing was switched from water and sewer being charged on the annual tax notice to separate bimonthly utility bills. This allows customers to monitor their water use and make changes immediately. In addition, with the data provided from the AMI system, staff were able to develop an equitable seasonal tiered water rate structure (Table A-8) that rewarded water wise and average users and encouraged high users to reduce. An analysis of the rate structure showed that most single-family residential customers, approximately 65 percent, stayed within the first tier during the summer (Table A-9). It was also found about 18 percent were in the second tier and 17 percent were in the third tier.

Table A-9 Breakdown of Accounts in Each Tier

Single-Family Residential				
	May/Jun	Jul/Aug	Sep/Oct	Nov/Dec
Tier 1 0–60 m^3 (0–21.2 hcf)	66%	65%	69%	72%
Tier 2 61–90 m^3 (21.5–31.8 hcf)	18%	18%	17%	17%
Tier 3 91 m^3 + (32.1 hcf +)	16%	17%	14%	11%
Multifamily Residential				
	May/Jun	Jul/Aug	Sep/Oct	Nov/Dec
Tier 1 0–60 m^3 (0–21.2 hcf)	90%	90%	91%	91%
Tier 2 61–90 m^3 (21.5–31.8 hcf)	6%	6%	6%	7%
Tier 3 91 m^3 + (32.1 hcf +)	4%	4%	3%	2%

Note: m^3 = cubic meters; hcf = hundred cubic feet

Most of the customers triggering the second or third tier fell into four different categories:

- Leaking services
- Secondary suites
- Large families
- Excessive irrigation

The median for single-family accounts was 45 m^3 (15.8 hcf) per bimonthly billing period (Figure A-6), and the median for multifamily accounts was 30 m^3 (10.6 hcf) per bimonthly billing period (Figure A-7).

While secondary suites and large families were taken into consideration and an additional allotment was recommended, the rate structure was converted back to a uniform structure due to political pressure. However, the frequent billing remains and water use awareness has improved considerably.

Enhancing Conservation Programs

Once customers have been motivated to conserve water through financial incentives and understand their consumption patterns, they become more interested in their water use. The advanced metering infrastructure provides hourly reading information to the nearest liter. The system's monitoring capabilities have already proven invaluable in early leak detection and in identifying customers with high water consumption. As of 2015, 3,000 leak letters have been sent out, and accounts that have a leak greater than 50 liters/hour (13 gallons/hour) are continuously flagged for review.

In addition to the benefits attributable to the AMI system, water savings were seen through a 40 percent increase in rebate applications for high-efficiency toilets and washing

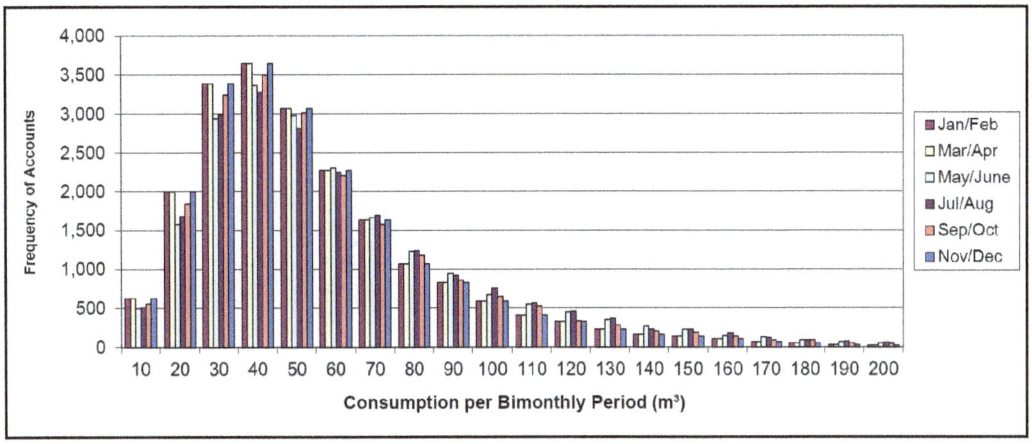

Figure A-6 Consumption frequency for single-family accounts

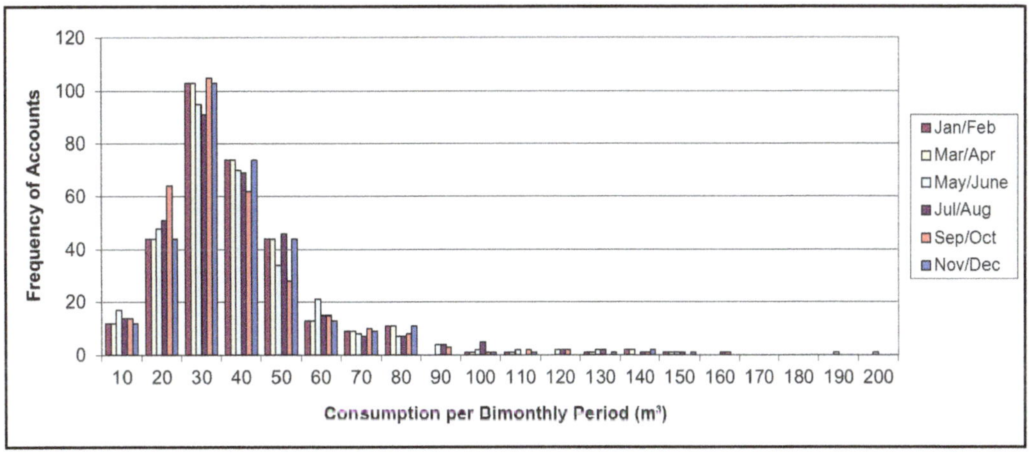

Figure A-7 Consumption frequency for multifamily accounts

machines, irrigation and landscape assessments, industrial, commercial and institutional audits, and purchasing of subsidized indoor and outdoor low-flow kits. Rainwater harvesting systems were installed at municipal facilities and several households. The success of these programs has helped defer future infrastructure development.

Evaluating the Benefits of Conservation

The AMI system also can demonstrate the benefits of conservation. For example, a two-week study was performed on 50 homes that installed high-efficiency toilets (HETs) and washing machines. The results from AMI data show on average a 30 percent reduction for replacing one 13-liter (3-gallon) or greater toilet with an HET and 19 percent for installing a high-efficiency washing machine (Figures A-8 and A-9).

AMI was also used to determine irrigation use for the irrigation assessments. On average a 68 percent water savings is seen. For example, the data from AMI showed that one household achieved a 79 percent reduction in irrigation water use after the assessment (Figure A-10). The assessment was done in mid-July 2011.

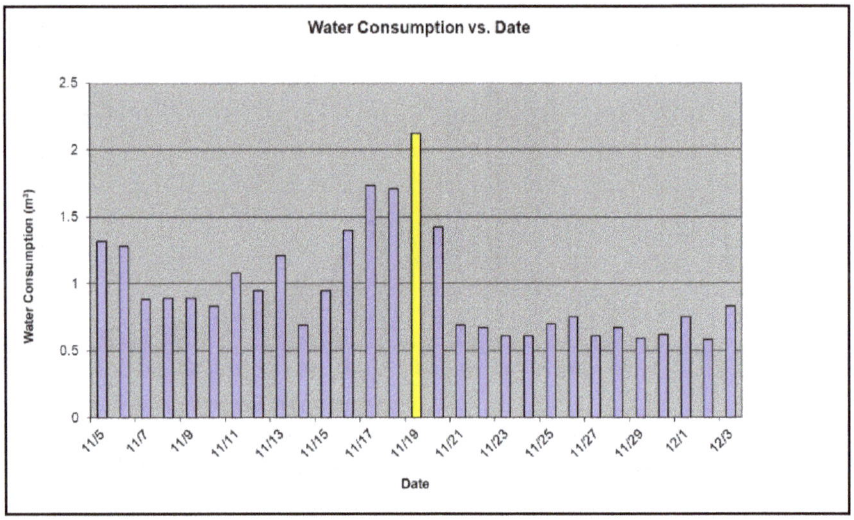

Figure A-8 Thirty percent reduction with a high-efficiency toilet (yellow bar indicates installation date)

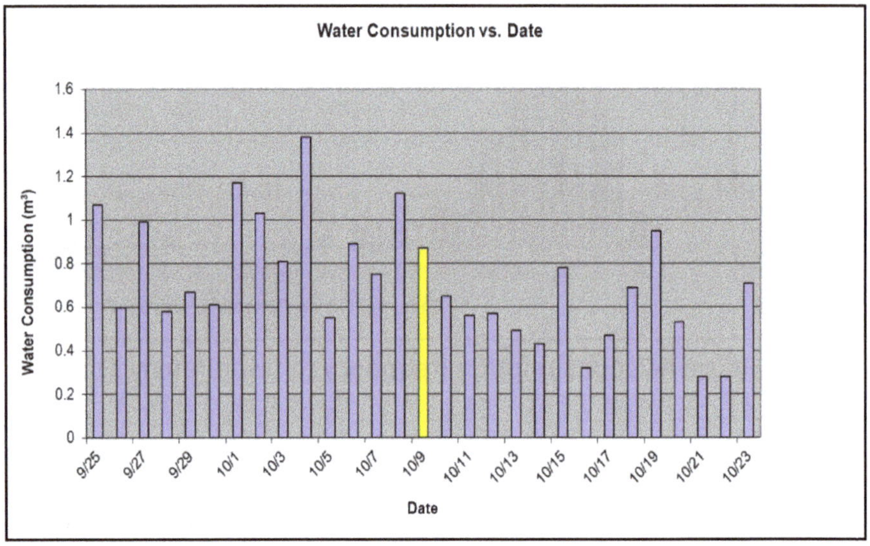

Figure A-9 Nineteen percent reduction with a high-efficiency washing machine (yellow bar indicates installation date)

Analyzing Demands for Future Programs

The AMI data can be analyzed to determine where water is being used inefficiently and helps target areas to address. Using data from Abbotsford's geographic information system and AMI, a residential consumption analysis was developed by mapping individual customer accounts. This allowed for a comparison of consumption during different times of the year, as shown in Figures A-11 and A-12.

Conclusion

Understanding the factors that influence water consumption patterns in distribution systems is fundamental for both long-term and short-term tasks. The AMI system has

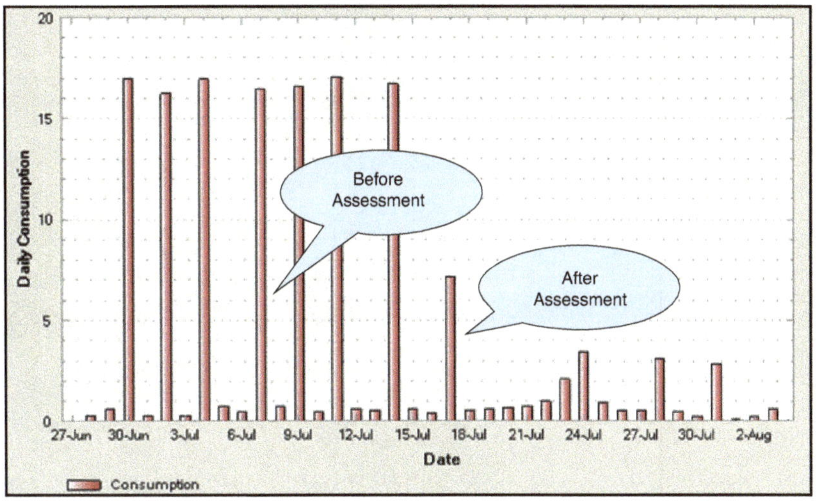

Figure A-10 Reduction in water use from an irrigation assessment

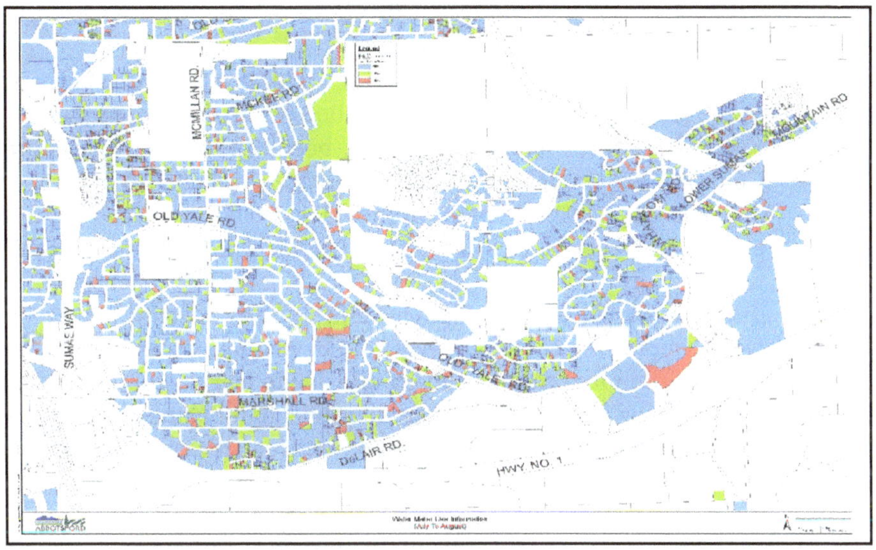

Figure A-11 Winter consumption in a heavily irrigated neighborhood (green and pink represent high water use)

played a significant role in helping to monitor demands and support conservation efforts in Abbotsford and will continue to play a significant role in planning for future demands.

CASE STUDY #5: SEATTLE, WASHINGTON – SAVING WATER PARTNERSHIP – WATER SMART TECHNOLOGY PROGRAM: INDUSTRIAL, COMMERCIAL, AND INSTITUTIONAL PROGRAM

Background

The Saving Water Partnership is a part of the Seattle Water Supply System Regional Water Conservation Program. A component of the Saving Water Partnership is the Water Smart

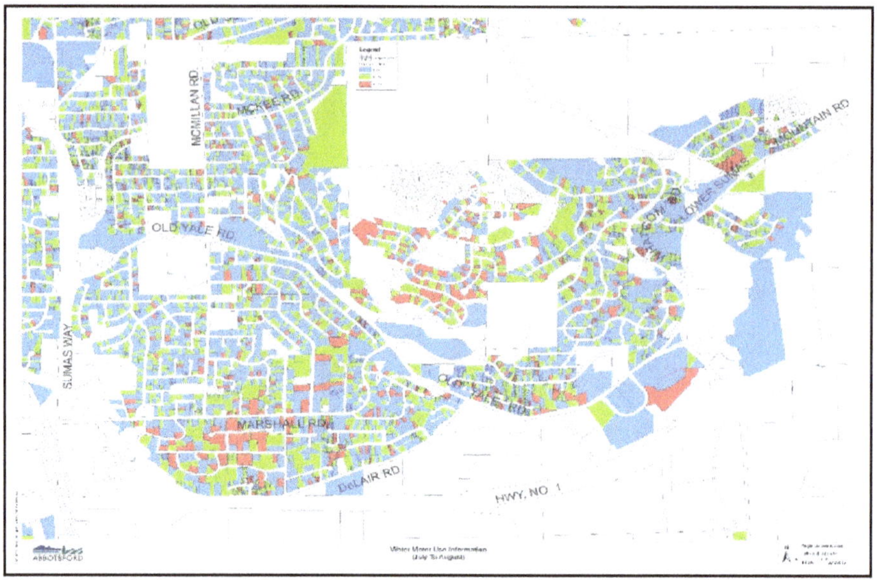

Figure A-12 Summer consumption in a heavily irrigated neighborhood (green and pink represent high water use)

Technology Program, which provides free technical assistance and financial incentives (e.g., rebates and cost shares) to reduce water use at commercial, industrial, and institutional facilities. The program has been operating with reported success since 1995. Conservation opportunities include replacing toilets and urinals, converting ice machines and refrigeration equipment from water cooled to air cooled, instituting other types of pass through cooling, installing high-efficiency commercial clothes washers and laundry water reclaim systems, replacing liquid ring medical vacuum and air compressors, introducing other medical equipment efficiencies, recycling and reusing industrial process water, improving cooling towers, and developing other water use efficiency technologies. Initially the program focused on midsize to large commercial customers. Since 2003, the program has included a special focus on small businesses in order to foster participation and capture savings from this sector, which historically had not shown significant participation. Emphasizing participation from small businesses also provides service equity and outreach to historically underserved customers and populations.

Program staff and consultants provide efficiency solutions through on-site assessments, technical review, product evaluation, workshops, and case studies. Program financial incentives provide standard rebates or custom rebates of up to 50 percent of the installed costs for projects determined to be cost-effective to the utility. From a customer perspective, most program participants have a simple payback of less than two years on their investment.

Objectives of This Program

The Water Smart Technology program supports the objectives of the regional Saving Water Partnership, which include:

- Helping customers use water wisely
- Ensuring the utilities are prepared to be resilient for curtailment events and future supply challenges from climate change
- Preserving customers' ethic of conservation as one element of stewarding water resources and the environment

Key Program Design Elements

- **Rebates vs. audits:** The Water Smart Technology program offers rebates for installed and completed projects. Only offering free water use audits to businesses is not a proven strategy to achieve conservation goals, because follow-through on implementing audit recommendations is not guaranteed. Offering rebates is often the significant factor to moving a business to implementation.
- **Require cost share:** Requiring at least 50 percent of the project cost to be supplied by the participant instills more ownership of the results and allows a program budget to be spread across more customer participants. Because the basis of the program is long-term utility savings, customer ownership and maintenance are key to long-term success.
- **Include non-rebate services:** Some business customers will implement conservation without utility rebates. Therefore, in addition to rebates, the program uses a variety of other activities to support and encourage customers to conserve. These include technical assistance, workshops, seminars, business recognition actions, product testing and research, written reports and technical manuals, and development of checklists to help with behavioral changes. Recent reports have included "How High-Efficiency Restroom Fixtures Really Perform" and "Optimization of Cooling Tower Water Treatment," as well as an reverse osmosis optimization checklist.
- **Persist:** Persistence is necessary with the nonresidential sector. It can be difficult to determine and reach the appropriate person within the customer's organization. Additionally, it can take several years for large businesses and institutions to budget their share of project costs and to justify projects to their decision makers.

Outreach Strategies

Program outreach and delivery focuses on four strategies.

1. **Vendor driven:** Equipment and service providers are the primary marketers for the program. They use the rebates as sales leverage as they interact with their customers. And because the rebated equipment comprises only the most water-efficient models, the potential for free riders from this marketing strategy is reduced.
2. **Partnerships:** The program partners with trade groups, energy utilities, agencies, and other service providers. Partnerships with energy utilities, when projects reduce both water and energy use, have leveraged water utility funding by increasing the overall rebate to the customer.
3. **Target select business categories or end uses:** Although it is offered to all nonresidential customers, the program has emphasized large customers such as hospitality, medical facilities, schools, and institutions. There has been a multiyear effort to help the University of Washington reduce overall water use. A resource efficiency committee meets every other month to go over the latest efficiency projects, review results, and discuss suggestions for new projects.
4. **Multiple-resource one-stop shopping:** Outreach, education, and technical assistance are provided to businesses to help them conserve resources, reduce or prevent pollution, and become more sustainable. Services are offered in five program areas: (1) water conservation, (2) waste prevention, (3) recycling, (4) stormwater pollution prevention, and (5) green building. This one-stop shopping provides an efficient approach to serving business customers and leverages the water conservation budget.

Budget and Staffing

Between 2013 and 2018, the Water Smart Technology Program will spend $4.8 million to meet the objectives of the Saving Water Partnership. The program is funded from both capital and operating budgets of participating utilities. Generally, program elements related to equipment rebates are funded from the capital budget, while program elements related to information and education are funded from the operating budget. This financing strategy reduces the program's impact and competition for operating budget funds, which can be limited in some years. The program is administered by approximately one full-time employee and the assistance of consultants to augment outreach and technical assistance capabilities.

Program Evaluation

The Water Smart Technology program is evaluated periodically to ensure that actual savings are achieved. The program is estimated to have saved 41.6 million gallons per year in 2012. Individual projects are evaluated prior to authorization for rebates. For some projects, such as coin-operated clothes washers, a standard rebate is provided based on measured savings from a large sample of customers. For other projects, a portable flowmeter or standard flowmeter is installed for pre- and post-evaluation. Site inspections are done for all large projects and many smaller ones. Long-term tracking is done from billing records and periodic checkups with facility managers and vendors. Payment and process procedures are checked by independent auditors to ensure accuracy.

Agency Summary

The Saving Water Partnership includes the City of Seattle (Seattle Public Utilities) and 18 wholesale water customers. Utilities in the Saving Water Partnership provide drinking water to 1.1 million people in the greater Seattle area. In 2012, Seattle Public Utilities provided 86 million gallons per day of water for its retail and wholesale customers in the Saving Water Partnership.

References and Contact

Water Smart Technology Program website:
www.savingwater.org/Businesses/index.htm

Resource Venture website:
www.seattle.gov/util/ForBusinesses/GreenYourBusiness/index.htm

CASE STUDY #6: FORT COLLINS, COLORADO – ASSESSMENTS LEAD TO CUSTOMER ACTION: SPRINKLER SYSTEM AUDIT PROGRAM – LANDSCAPE

Fort Collins Utilities believes that assessments for customers lead to behavior change and retrofits that improve water efficiency. Since 1999, Fort Collins Utilities has administered a sprinkler system audit program as part of its water conservation efforts. The program seeks to inform and educate the community, one household at a time, about how to run sprinkler systems efficiently and reduce excessive water use

The free audits are offered to single-family homes and homeowners associations (HOAs). Analyzing audit participants' water use consistently finds that they save an

Figure A-13 Sprinkler head with catch cans

average of 20 percent of their outdoor water use. In 2013, the audit participants saved almost 2 million gallons.

An Evolving Program

During the first year of the program one auditor performed 60 home audits; only basic training was provided, large bowls were used as catch cans, and handouts were minimal. Believing in continuous improvement, Fort Collins Utilities made program changes including training, forms, reports, scheduling, and handouts. Now during a typical season, up to five auditors complete 350–450 home and 10–15 HOA audits. Over the past 13 years, 3,434 homes and 101 HOAs have received audits.

Home Audits

For a residential audit, auditors meet one on one with customers at their homes for up to two hours. The resident who operates the sprinkler system typically participates in the audit to learn about their system and see firsthand any maintenance needs. They learn how the lawn uses water, how to program the controller, and the importance of periodic system checks. The auditor prepares a report detailing recommendations for scheduling changes, maintenance needs, and system upgrades. Information about available sprinkler equipment rebates is also provided.

An auditor will:

- Inspect each zone to identify maintenance needs
- Perform catch can tests to measure the application rate and how evenly water is being applied
- Measure system pressure and grass root depth
- Develop a custom watering schedule
- Share the results and recommendations with the homeowner

Figure A-13 shows how sprinkler head efficiency is tested using catch cans. Figure A-14 pictures a homeowner being educated about her irrigation management system by an auditor.

Figure A-14 Auditor educating a homeowner

Homeowner Association (HOA) Audits

HOA representatives and the landscape maintenance contractor accompany the auditor for an up-to-four-hour audit of common-area sprinkler irrigation systems. Each zone is visually evaluated for any leaks, broken or misaligned sprinkler heads, and poor coverage. Catch can tests are not performed for these large properties, as the time is better spent identifying maintenance issues. The audit report includes a zone-by-zone evaluation, photos, explanation of problems, watering schedule, and a list of recommended maintenance tasks by priority. Because recommendations for large properties may be expensive, the report is typically presented at an HOA board meeting so the suggestions can be explained and questions answered.

Staffing and Training

Utilities staff recruit, hire, and supervise four or five seasonal auditors who work from mid-May to mid-September. Figure A-15 pictures an auditor training. Many of those hired are college students as the summer schedule works well for them; some return for multiple summers. Auditing experience isn't necessary, but good communication skills are mandatory. In addition, a program administrator is hired to provide scheduling, keep supplies available, and send the audit report and an evaluation to participants.

A comprehensive three-day training includes the basics of sprinkler systems, hands-on practice with controllers, and role-playing audits. Colorado State University professors provide training about soils and turfgrass and water use. An important part of the training is how to communicate with customers in a way to effectively educate the city's diverse population. The first week of audits, new auditors team up in the field with seasoned auditors. First, the experienced auditor models an effective audit. Then it's the new auditor's turn to conduct the audit with coaching from the experienced auditor.

Throughout the summer, staff and the auditors meet biweekly to discuss issues and questions that arise, share stories from the field, and receive more advanced training.

Figure A-15 Auditor training

Figure A-16 Catch can test

Figures A-16 through A-19 present various auditors conducting catch can irrigation sprinkler text analysis.

Scheduling

Appointments for home audits are scheduled Monday through Friday for two hours at 8:30 a.m., 10:30 a.m., 1 p.m., and 3 p.m. Late-day or Saturday appointments are also accommodated. Auditors work alone and perform four audits a day with a half-hour for lunch. HOA audits are scheduled with two auditors for four hours.

Customers request an audit online or by voice mail. An online scheduling program is used to track customer information, appointments for each auditor, and completed audits. Participants receive automated appointment confirmation and reminder emails. One challenge is connecting with people to make an appointment. To help with this, the program's self-scheduling feature will be used in the future. After audits are scheduled, they are juggled among auditors to keep appointments as close together as possible.

Figure A-17 Catch can test analysis

Figure A-18 Auditor sharing results

Customer Feedback

An online evaluation is sent to some participants via email; others receive a paper copy in the mail. In 2013, 84 percent of participants said the audit exceeded their expectations. This percentage has significantly increased over the past few years as more time has been dedicated to training and a staff irrigation professional provides in-depth expertise and support to the auditors.

When asked if they had made changes since the audit, 93 percent of participants responded that they had. The most common changes were adjusting sprinkler heads, changing the watering schedule, and fixing leaks. All comments are taken seriously and guide future program improvements.

Figure A-19 Auditor, catch cans, and van

Water Use Tracking

Utilities uses a software tool to track program and rebate participation and resulting water savings. The software uses customer account records, geographic information system (GIS) data, tax assessor information, and current/historic weather data. An analysis of 671 sprinkler audit participants showed an average annual water savings of 12,110 gallons and a median water savings of 7,690 gallons, about 20 percent of their outdoor use. These data indicate that customers are continuing to save water.

Budget

The program budget is allocated through the city's biennial budget cycle. Annual costs run between $17,000 and $18,000 per year, including equipment, training, publicity, mileage, and labor. The cost per audit typically ranges between $40 and $50.

Annual Planning Schedule

- February: begin auditor recruitment
- April: complete hiring
- April and May: finalize training schedule, materials, and equipment
- End of May: three-day training
- June through August: perform audits, ongoing scheduling
- September: compile evaluations and write report

Agency Summary

Fort Collins was a community of 144,000 residents when this evaluation was done. Located on the plains just east of the Rocky Mountains, Fort Collins Utilities provides water, wastewater, stormwater, and electric services to the community. In 2012, it served 8.7 billion gallons to 33,800 water service connections.

CASE STUDY #7: COBB COUNTY, GEORGIA

Introduction

In 2006 Cobb County, like much of Georgia, was experiencing a very hot and dry weather pattern. Our per capita water use during peak season was up 20 percent above previous years. At the same time Cobb County Water System (CCWS) adopted a new rate structure, to comply with the Mandatory Water Conservation and Supply Plan for the Metropolitan North Georgia Water Planning District. The mandate required that all 65 utilities in the water district adopt a tiered conservation rate structure targeting discretionary water use, particularly outdoor water use, by January 2006. Cobb County's previous rate structure had been a flat volumetric rate with a small surcharge for usage in excess of 150 percent of average winter use (December, January, and February); additionally, the rate structure capped wastewater charges at 150 percent above the winter average. During Cobb's rate review it was determined that wastewater costs were being subsidized by water charges and that capping sewer was a conservation disincentive, because the small surcharge and low water rate sent no price signal on discretionary usage.

As part of the new rates CCWS began billing wastewater equal to metered water use to recoup the extensive costs of the conveyance system. The new three-tiered rate along with the removal of the sewer capping program raised rates for the system's highest users by more than 100 percent. The rate structure was the first step in attempting to reduce outdoor water use, but 2006 showed the highest seasonal use despite the rate structure change. The high seasonal use and the newly increased rates had caused some Cobb County Water System customers to oppose the rate structure. CCWS wanted to reduce outdoor water use and increase the quality of customers' landscapes while reducing their water bills. Though the rate structure alone did not reduce the use for our highest users it did successfully get their attention which helped us launch our first outdoor water use program targeted at reducing peak summer use.

Program Design

The first step of the program was to determine what key concepts we wanted program participants to understand. Before developing the program CCWS met with landscape professional, irrigation professionals, University of Georgia Urban Agriculture Department, and our UGA Extension agent. Because no program can address all aspects of outdoor water use the program was crafted around three key concepts that were developed from our work with industry professionals:

1. The number one killer of plants in Georgia is overwatering. This came from our work with UGA Extension that, with UGA Center for Urban Agriculture, had been cataloging plant mortality. The disease patterns and records they had indicated that the overwatering of plants was either the direct cause or a contributing cause that made an environment hospitable to disease.
2. Most landscape plants need about one inch of water applied once a week in our area. This came from research by UGA on the Piedmont ecoregion and was confirmed by our irrigation and landscape experts.
3. Your irrigation system should be programmed to deliver this amount of water to the plants in the absence of rain. The methodology and tips on how to do this were created with the Georgia Irrigation Association.

Figure A-20 Give Them an Inch . . . Grow a Yard program logo

The key concepts were simple, and additional information and tools helped customers achieve more attractive and water efficient landscapes: an at-home water audit brochure, outdoor kits, and the Water: Use It Wisely Southeastern Watering Guide. The outdoor kit contained soil moisture probe, hose nozzle shutoff, and a rain gauge. To drive home our key concepts, the program name and logo were developed and trademarked. Figure A-20 pictures the Give Them an Inch . . . Grow a Yard program logo.

The program was completely free to participants and had two levels of participation. The Give Them an Inch Communities was a residential program where Cobb Water came to local neighborhoods or homeowners associations (see program logo in Figure A-20). We gave a presentation on outdoor water efficiency and how to calculate irrigation system output. We also brought the program materials for every resident of the community. All designated communities received yard signs for their neighborhood entrances. The Give Them an Inch Partners was the professional arm of the program. Landscaping and irrigation companies and nurseriescould participate by carrying or providing our materials to customers. We provided the materials to the partners at no charge, as well as window clings for their businesses and trucks. The companies were also designated on our website.

Program Results

Cobb Water was pleased with the program results. We had an original goal of reducing per capita use during peak season by 5 percent. In 2007, though the drought had intensified, per capita water use prior to the mandatory ban issued in October of 2007 was down on average 8 percent. The program has continued, and peak use has continued to be below the numbers experienced in 2006.

The program was originally developed to respond to peak use that was driving our wholesaler's use above its permitted capacity. CCWS purchases all of its water wholesale from Cobb County–Marietta Water Authority, and our system represents 67 percent of their total sales. A few things happened to magnify the impact of the program. The interactions with our communities and the garden clubs, where we introduced the program, demonstrated that we had a significant number of customers who were not only watering more than an inch a week, they were watering incorrectly for the clay/loam soils found in CCWS's service area. When the program was developed, all Georgia citizens were limited to three mandatory watering days. Our surveys of program participants and our discussions with program attendees, as well as post-program follow-ups, showed over 70 percent of program participants were watering all three days, regardless of rainfall or plant need. Not only were they watering three days, they were allowing their irrigation systems to water for 20–40 minutes per zone continuously. In the Piedmont ecoregion, the ideal absorption window is about 6–8 minutes. The irrigation audit created for the program not only instructed the citizens on how to calculate system output and improve distribution uniformity with simple adjustments, it also discussed how best to schedule irrigation to increase the depth of absorption and minimize runoff using a cycle-and-soak methodology.

The total water ban issued in October 2007 and lifted in June 2009 also helped to drive home the message that plants are sturdier and more likely to survive hardship if they have been established using the techniques just described to encourage strong deep root growth and reduce frequent water dependency. When the ban was lifted, many customers saw established turf that had gone dormant in the absence of irrigation come back after rain and supplemental irrigation returned. This initial program and the continued use of the materials as well as supplemental trainings have had a lasting impact on CCWS's peak usage. From 2010 to 2012, the region again experienced a serious drought. In the 2005–2007 drought, seasonal peak usage was between 35 and 45 percent; during the 2010–2012 drought, the system's seasonal peak usage was between 22 and 25 percent.

Cobb Water considers our partnership with professional green industry groups, which provided technical information for our irrigation audit brochure and support for the program, to be a key to the program's success. Some customers will listen to the government for their information and some will listen to their landscapers, so a secondary benefit of this program is that citizens were hearing a consistent message from both. Another unexpected finding when we first surveyed our program participants was that they had no interaction with their landscape or irrigation professionals. The landscapers came to the home when the citizens were at work, and the two never discussed the system or the scheduling. The at-home irrigation audit also encouraged citizens to hire certified irrigation professionals and have their systems checked at least every three years by a certified irrigation auditor. The program helped to provide citizens with information so they could begin to have some ownership and understanding of their irrigation system and feel comfortable talking to their landscape professionals.

Building on the Success

Since establishing the Give Them an Inch . . . Grow a Yard program, Cobb Water has built upon its relationships with irrigation professionals and other landscape experts to expand outdoor programs. Recognizing that the field-level "boots on the ground" landscapers were not offered training and were actually wasting more water and damaging water quality with their practices, the water efficiency program partnered with UGA Extension and Cobb County Stormwater Division to create The Sustainable Landscape Training. This program was based on a landscape and irrigation tool kit that was purchased by Cobb County Water System. Through hands-on activities, local professionals in landscape and irrigation taught the attendees how to use the tools in the kit to better manage irrigation systems. Cobb Water closed the training by providing an overview of other training opportunities and certifications available to green industry professionals. The class had no texts and very little lecture. After the class the attendees got to keep their kits, with approximately $200 in tools with them to use in the field. They were also listed on our website as having completed the training.

CCWS worked with UGA Extension to follow up with the attendees using a survey instrument. The class was taught in the winter when the industry has downtime, and the survey was sent out midway through the busy season. We had a response rate of 42 percent. The questions focused on whether the attendees were using the tools and asked a couple of questions about sustainable best practices covered during the class. For example, one big issue was what to do with grass clippings, so a question about debris management asked about disposal into storm drains versus use of grasscycling (leaving clippings on the lawn). Respondents indicated that debris in storm drains contributes to flooding and that in some situations they were removing the bag from mowers and incorporating grasscycling.

CCWS offered this class for two years and trained 55 professionals. Due to the high cost of the class and the loss of some key instructors, CCWS does not currently offer the class.

In 2013, the US Environmental Protection Agency program WaterSense® launched Sprinkler Spruce-Up. Cobb Water had been struggling with the difficulties customers and professionals had with scheduling their controllers. To celebrate Sprinkler Spruce-Up, CCWS purchased eight irrigation controllers typical of what is found in the service areaincluding state-of-the-art WaterSense–labeled controllers and basic manual controllers. CCWS again partnered with irrigation experts and a held a class, Correcting Common Mistakes in the Field, during Sprinkler Spruce-Up in May. The irrigation professionals used the controllers to teach hands-on scheduling and then moved to the on-site irrigation system to teach how to make basic repairs.

The class was attended by 60 professionals, many of whom work in the municipal government field managing parks and irrigation at local universities. This made it easy to follow up with a portion of attendees. As usual a survey was sent to attendees, and about 39 percent responded. Respondents indicated that the hands-on training is the most beneficial, because they can immediately use it in the field. A few participants said they were able to more effectively troubleshoot high water bill questions by observing the customer's irrigation system and checking the controller. This indicated to us that the course successfully addressed core competencies. Additionally, a small percentage of attendees now felt they could make the case more successfully with their organizations for incorporating WaterSense–labeled controllers.

The investment in the controllers has been worthwhile. During Smart Irrigation Month, CCWS displays a WaterSense–labeled controller in the customer service office lobby with instructions so some of the 50 customers who visit CCWS during the day can practice setting up and scheduling a controller. They can also call back for a member of the water efficiency staff to come up and work with them. Water efficiency staff worked directly with two dozen customers in 2014, and the front office staff said they received approximately 10–15 requests a week from customers asking if they can use the controller.

The controllers were also used for a landscape and irrigation open house during Smart Irrigation Month in 2014. CCWS and UGA Extension partnered to offer office hours on a Saturday in July. Customers could drop in and ask their irrigation and landscape questions. CCWS had the controllers so customers could practice scheduling; there were also sprinkler nozzles and spray bodies and assorted components of drip irrigation. The open house format allowed the citizens to control the time they came and how long they stayed. Because all the materials were set up in stations, customers interested in irrigation could actually get questions answered and practice some hands-on skills like changing a sprinkler head or scheduling a controller.

CCWS continues to work with garden clubs and civic groups to present irrigation and water efficient landscaping to citizens. One of the most popular is the elementary school fall planting tour. CCWS staff and the WaterSmart program choose two economically disadvantaged schools and provide all the materials to do a water efficient planting. The students learn the principle of "right plant, right place," best planting practices, and that fall is the optimum time for planting many plants in Georgia since they can establish a strong root system over the winter.

Outdoor water use continues to be a focus for Cobb County, and different programs resonate with different audiences. Since 2006 CCWS's annual average daily demand has dropped from 67 mgd to 57 mgd in spite of an increasing population, and part of that success has been a reduction in outdoor water use.

CASE STUDY #8: NEW YORK – NEW YORK CITY CHALLENGES HOTELS TO SAVE WATER

When you think about water utilities that have to implement water conservation programs to prepare for drought or other water shortages, New York City is probably not one of the first that comes to mind. Yet the city has experienced nine periods of drought over the past 75 years (NYC DEP 2017), and can be challenged during the summer when demand peaks due to outdoor watering.

The New York City Department of Environmental Protection (DEP) manages New York City's water supply, providing more than one billion gallons of water each day to more than nine million residents, including eight million in New York City. The water is delivered from a watershed that extends more than 125 miles from the city, comprising 19 reservoirs and three controlled lakes. Approximately 7,000 miles of water mains, tunnels, and aqueducts bring water to homes and businesses throughout the five boroughs, and 7,500 miles of sewer lines and 96 pump stations take wastewater to 14 in-city treatment plants.

The water conservation challenge is one part of New York City's Water for the Future program, which will help to ensure clean, reliable, and safe drinking water for nine million New Yorkers for decades to come. The central part of the program is repairing the Delaware Aqueduct, which supplies roughly half of the city's daily drinking water needs. Ground has been broken on two 800-foot-deep shafts and a 2.5-mile bypass tunnel around a portion of the aqueduct that is leaking. To facilitate these repairs to the aqueduct, the tunnel must be temporarily shut down between 2021 and 2022. In preparation for the shutdown, DEP has developed a combination of conservation programs and supplemental supplies that will ensure an uninterrupted supply of water. The program was initially estimated to cost more than $2 billion, but through advances in the engineering and design of the bypass and the water supply projects to support the repair, the estimated cost has been reduced to approximately $1.5 billion.

New York City's Conservation Planning

To help reduce demand, the DEP is carrying out a concentrated effort to reduce citywide water consumption by 5 percent. DEP is pursuing six strategies to meet its goal – (1) a municipal water efficiency program focusing on city-owned facilities, (2) a residential water efficiency program, (3) a nonresidential water efficiency program, (4) a water distribution and system optimization program that includes leak detection and management, (5) a water supply shortage management program that will outline the rules and triggers called for when a shortage condition occurs, and (6) a wholesale customer demand management program that will assist the 10 largest upstate wholesale customers in development of their own water demand management plans.

For the nonresidential program, DEP is looking to increase its engagement with the city's commercial sector, which makes up approximately 20 percent of total water usage ("Water Demand Management Plan," NYC DEP2013). In 2009, New York City enacted new requirements for large buildings to annually measure and report their energy and water consumption in order to allow for benchmarking. With this new reporting, building owners have a new level of understanding with respect to their use of energy and water. DEP

Table A-10 Elements of the Mayor's Water Challenge to Hotels

Tracking water use	To accurately measure water consumption, all Challenge participants must have automatic meter readers (AMRs) installed in all occupied spaces larger than 10,000 square feet. All water usage will be tracked through AMR.
Water reduction strategy	All participants will be required to develop a water conservation plan that maps out a strategy to reduce water usage to meet the Challenge goal. This plan must be submitted within the first three months of the challenge.
Regular partner meetings	The DEP, in partnership with the Mayor's Office, will organize regular meetings with all Challenge participants to provide updates on the Challenge and encourage the exchange of ideas. At least one representative from each hotel should plan to attend each meeting.
One-on-one meetings	DEP and Mayor's Office staff and the Challenge participants will meet once to review the participants' progress, consider any obstacles to achieving reductions, and discuss a strategy to reach the water reduction goal.
Working groups	DEP will convene voluntary working groups as needed to help develop new initiatives and/or to resolve issues as they arise.
Promotion and recognition	DEP and the Mayor's Office will provide promotion and recognition of the Challenge participants including press conferences, news releases, and information posted on the PlaNYC and DEP websites.

Source: NYC DEP n.d.

is taking advantage of this new awareness to engage sectors on their opportunities for reducing water use.

The Mayor's Challenge

In 2013, the DEP partnered with the Hotel Association of NYC and the Mayor's Office to develop the New York City Water Challenge to Hotels. Participating hotels were asked to voluntarily reduce their water use by 5 percent. Eleven hotels took the Challenge, agreeing to track their water use and develop a water conservation plan. The other elements of the Challenge are outlined in Table A-10. For partner meetings, the DEP brought in a number of external speakers to share information about best management practices the hotels could undertake as a means to fast-track savings opportunities.

Each hotel received a monthly progress report (Figure A-21) to show it where it stood in consumption for the month and year as compared to the previous year. To get the competitive juices flowing, the report also included a graph to show where the hotel's monthly water use intensity (water use per square foot [water use/sq ft]) compared to that of the other hotels participating in the Challenge, as well as the monthly average for all participating hotels.

Before the Challenge began, the participating hotels had consumed between 50,000 and 320,000 gallons of water per day during the previous 12 months. NYC DEP estimated they could realize savings of between $10,000 and $70,000 in annual water and sewer costs if the 5 percent reduction goal was met ("Water Demand Management Plan," NYC DEP 2013).

The Sheraton Tribeca achieved the best results, reducing its annual consumption by more than 20 percent. Three other hotels, the InterContinental New York Barclay, the Ritz-Carlton, and the Carlton Hotel, also surpassed the 5 percent goal, each reducing consumption by more than 10 percent. In total, the four hotels that achieved at least a 5 percent reduction in water consumption conserved nearly 11.3 million gallons of water over the course of the Challenge. The commissioner gave the New Yorker hotel an honorable mention for managing to maintain the lowest water use per square foot for the majority of the Challenge.

Figure A-22 pictures the recipients who received recognition in the 2014 Mayor's Water Challenge to Hotels.

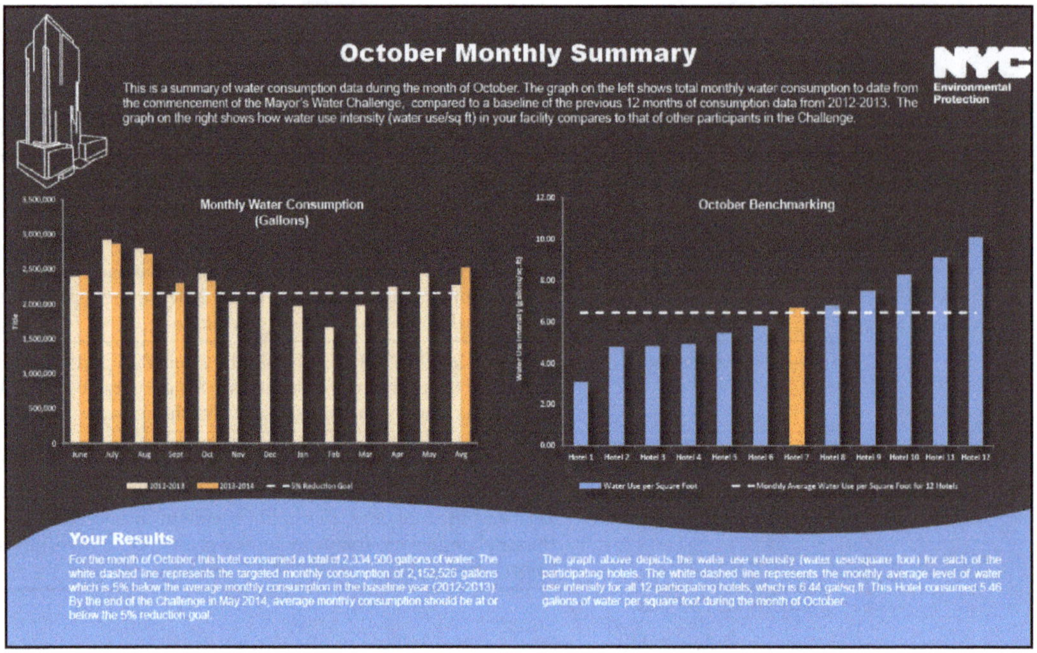

Figure A-21 Elements of the Mayor's Water Challenge to Hotels
Source: NYC DEP2013

Figure A-22 Recipients of recognition from the 2014 Mayor's Water Challenge to Hotels

At the awards conference DEP also introduced its new "Hotel Manager's Guide to Water Efficiency," which provides tips on conserving water in guest rooms, common areas, food service areas, laundry, and pool operations, and on leak detection. DEP plans to distribute the guide to hotels throughout the five boroughs in an effort to replicate the success of the Mayor's Water Challenge to Hotels. The guide is available online at http://www.nyc.gov/html/dep/pdf/conservation/housekeeping-booklet.pdf.

Lessons Learned and Looking Forward

According to Vlada Kenniff, managing director for demand management and resiliency at DEP, one of the biggest lessons DEP learned was that working with a specific sector on water conservation is very important. "Uses and challenges vary from sector to sector, and it was extremely valuable to sit down with operators of hotels to talk about things that are specific to their institutions," said Kenniff. "We also learned that by not being overly prescriptive in the strategies our hotels took, we were able to see measurable results. Some hotels chose to run a marketing campaign while other hotels did retrofits, but all resulted in savings of 10 to 20 percent."

This public-private partnership challenged a select group of thirty NYC restaurants to reduce their annual water. DEP concluded its Water Challenge to Restaurants in September of 2015. This challenge was successful with 10 of the 30 restaurants being able to reduce consumption by at least 5%. Together, all of the winning restaurants saved 2.6 million gallons in New York City's 2015 water consumption." Page 14 of http://www.nyc.gov/html/dep/pdf/conservation/water_conservation_report2016.pdf

One change made for the new Challenge was to redesign the series of workshops to align with the water management planning steps included in *WaterSense at Work: Best Management Practices for Commercial and Institutional Facilities* (USEPA 2012). This provided for a consistent framework that participants could use and helped point to the WaterSense guide as a good resource to use as they carried out their efforts.

Looking forward, DEP expects to continue to use the Mayor's Water Challenge as a means to engage with other types of commercial facilities such as universities, hospitals, and office buildings.

References

New York City Department of Environmental Protection (NYC DEP). 2013. "Water Demand Management Plan." New York: NYC DEP.

New York City Department of Environmental Protection (NYC DEP). 2017. "Water Supply Shortage Management." http://www.nyc.gov/html/dep/html/ways_to_save_water/water-supply-shortage-management.shtml, accessed March 6, 2017.

http://www.nyc.gov/html/dep/pdf/conservation/water-demand-management-plan-single-page.pdf, accessed March 6, 2017.

New York City Department of Environmental Protection (NYC DEP). 2013 "Mayor's Water Challenge to Hotels Program Design." http://www.hanyc.org/sustainable-hospitality/wp-content/uploads/2012/06/Mayors-Water-Challenge.pdf, accessed March 6, 2017.

US Environmental Protection Agency (USEPA). 2012. *WaterSense at Work: Best Management Practices for Commercial and Institutional Facilities*. Washington, DC: USEPA.

CASE STUDY #9: GUELPH, ONTARIO – GRAYWATER REUSE AND RAINWATER HARVESTING REBATE PROGRAMS

Guelph, Ontario, is one of Canada's largest communities to rely on a limited groundwater source to meet the municipality's current and future water supply needs. In 2006 Guelph City Council endorsed the Water Supply Master Plan (WSMP). This detailed master plan evaluates the water demand associated with projected growth over a 50-year planning horizon as well as new water supply alternatives necessary to facilitate future community growth. The WSMP identified water conservation and efficiency as the most cost-effective and immediately available source of new water supply and, as such, ranked water conservation and efficiency as the primary water supply priority action. In support of this direction, the WSMP identified three time-based water reduction goals based on 2006 average daily water production volumes:

- Reduction of 10 percent (5,300 m^3/day or 1.4 mgd) in average-day water use by 2010
- Reduction of 15 percent (7,950 m^3/day or 2.1 mgd) in average-day water use by 2017
- Reduction of 20 percent (10,600 m^3/day or 2.8 mgd) in average-day water use by 2025

Further to the goals of the WSMP, the city's 2007 Community Energy Initiative set a sustainability performance goal of using less water and energy per capita than any comparable Canadian city. Together, these goals guide Guelph's current water conservation activities and bring greater emphasis to the relationship among water supply, wastewater treatment, and energy demand.

To achieve these goals, the city updated its Water Conservation and Efficiency Strategy (WCES) in 2008. This award-winning 10-year strategy was endorsed by Guelph City Council in May 2009 and identifies the preferred program, policy, and resource recommendations to achieve a further reduction of 8,773 m^3/day (2.3 mgd) by 2019, as well as various goals of the Water Supply Master Plan, Water and Wastewater Master Servicing Study, Wastewater Treatment Master Plan, Community Energy Initiative, and Council's Strategic Plan.

Through development of the WCES it was noted that many long-standing local municipal water conservation programs, such as toilet or clothes washer retrofit rebates, would reach saturation (meaning a large percentage of customers would already have retrofitted their fixture or device) within the 10-year planning horizon of the strategy. To meet the long-term reduction targets of the WSMP, further development was required in the area of research and evaluation of new demand management alternatives. As part of public consultation completed through development of the WCES, strong public and political support for decentralized demand substitution approaches – such as graywater reuse and rainwater harvesting – were expressed. Guelph City Council approved pilot programs for home-based graywater reuse and rainwater harvesting programs as part of the final WCES. The pilots allowed the city to investigate these alternatives and build the necessary technical and social capacity within the community to ensure future readiness for them.

In May 2009, the city of Guelph initiated the Residential Greywater Reuse Field Test to assess the feasibility of large-scale adoption of home-based graywater reuse technologies. During the study, a total of 25 participants installed home graywater reuse systems.

To promote uptake of graywater reuse technology, Guelph offered an incentive of Can$1,500 to homeowners installing an approved graywater system in either new or existing homes. Program participants were also provided with backflow prevention (premise isolation) devices at no charge and received financial compensation toward the required annual testing of these devices over a five-year period. In exchange for these incentives, participants agreed to allow city representatives to monitor the water quality of the graywater produced by their systems on a monthly basis for a period of 12 months, with a single final water quality sample taken 24 months after system installation. Additionally, participants were requested to participate in social feedback forums, interviews, and surveys to share their experiences and attitudes toward the technology.

The Residential Greywater Reuse Field Test project provided significant insight into the challenges and opportunities for the adoption of residential water reuse technology. This insight has also led to further research in partnership with the University of Guelph, local home builders, and technology providers to develop appropriate technologies and public acceptance strategies to encourage residential water reuse and recycling practices. This included improving treatment opportunities to produce effluent of higher aesthetic quality, minimizing maintenance requirements to system owners, and offering more cost-effective alternatives in residential settings.

Since the completion of the field study, the Resident Greywater Reuse Program continues to offer City of Guelph residents who install eligible graywater reuse systems with a Can$1,000 rebate. Since toilet flushing represents 25 to 30 percent of daily water use in the home, using reclaimed graywater from showers and baths eliminates the use of potable water for toilet flushing, resulting in measurable annual water and wastewater cost savings to those installing these systems. For more information on this program, or to download a copy of the field test report, visit https://guelph.ca/greywater.

In alignment with Guelph's Water Conservation and Efficiency Strategy, the city continues to be active in advancing rainwater harvesting adoption. The rainwater harvesting system rebate program provides qualifying homeowners and local businesses who install approved rainwater harvesting systems with a Can$2,000 rebate. Rainwater harvesting systems are individually designed and engineered for each installation. To qualify for the rebate, systems are required to meet standards of the Ontario Building Code as well as city rainwater harvesting system design standards. For more information on this program, visit https://guelph.ca/rainwater.

To learn more about the Guelph's water conservation programs and resources, visit http://guelph.ca/living/environment/water/.

CASE STUDY #10: CITY OF AUSTIN, TEXAS – MATURE CONSERVATION PROGRAM SHIFTS FOCUS FROM REBATES TO CODES DUE TO SATURATION

Austin Water provided about 35 billion gallons of safe, reliable, and high-quality water to more than 900,000 customers through more than 221,000 connections in the October 2014 through September 2015 fiscal year (FY) while also incorporating sustainable practices to protect the environment and help maintain quality of life for Austin residents. Water supply challenges in nondrought years are generally not as significant in Austin as in many other cities due to a contract with the Lower Colorado River Authority to provide a reliable "firm" water supply in an amount more than twice Austin's current demand. Nevertheless, Austin has historically embraced water conservation as a core value and kept long-term sustainability in mind as it set aggressive conservation goals. Addressing water conservation is especially important as Austin's population experiences rapid and sustained growth.

Austin's successful water conservation program, which in FY 2015 was credited with saving more than 8.2 billion gallons because of conservation activities and more than 2.7 billion gallons because of water reuse, includes measures beyond those typically found in many other Texas cities. This focus on conservation has proven effective, as exhibited in Figure A-23 showing Austin's downward trend in gallons per capita per day (gpcd) water use, which is calculated by dividing total water pumped from the treatment plants annually by the retail and wholesale service area populations. Figure A-23 also shows a similar downward trend in water use by residential customers.

Conservation Goals

Austin began its water efficiency program in 1983 with an ordinance that allowed the city to implement temporary water use restrictions in response to supply limitations or at times when peak demand approached treatment capacity. Over time, the focus shifted from viewing conservation mainly as a crisis response tool to seeing it as a means of extending the available water supply, reducing greenhouse gas emissions, and extending infrastructure capacity in the face of population growth and resulting development. As a

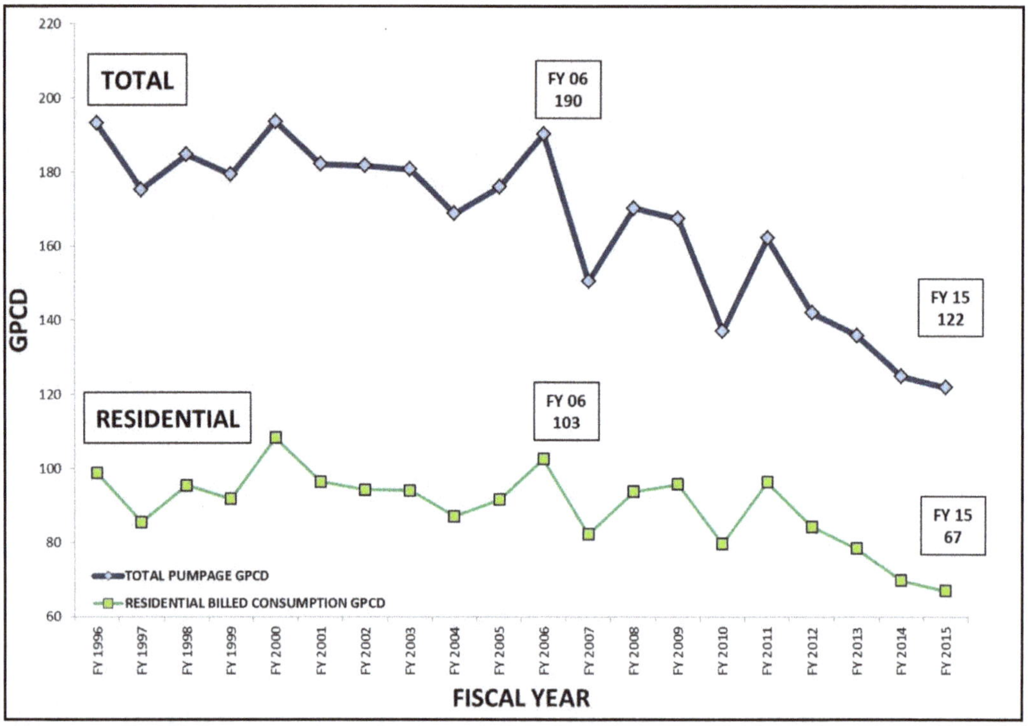

Figure A-23 Annual gallons per capita per day usage

result, Austin's conservation program has expanded to include a wide range of additional measures and strategies to reduce water use and enforce water use restrictions. Austin continues to look for ways to increase water savings, especially with sustained annual population growth of more than 2 percent (making it one of the fastest-growing cities in the United States) and a historic drought placing additional demand on available water supplies. Current water conservation goals include:

- Reduce peak daily water demand by one percent per year over a 10-year period ending in 2017
- Reduce average per capita water use to no more than 140 gpcd by 2020
- Delay the annual average use of 201,000 acre-feet of water for two consecutive years to avoid triggering additional payments to the Lower Colorado River Authority for water used above 150,000 acre-feet per year

Austin's water conservation program encompasses a wide range of measures and strategies to reduce water use including regulatory measures, conservation rates and pricing structures, incentives to install water efficient equipment, services to reduce water demand, promotion of auxiliary water sources, expansion of the reclaimed water system and increased availability of reclaimed bulk filling stations, and comprehensive customer education and outreach. In 2016, Austin's program received the highest score on the Texas Water Conservation Scorecard. Austin Water continues to consider using a variety of measures with adequate flexibility to allow for modification, expansion, or termination to accommodate new conservation goals or take into account new efficiency standards, market conditions, studies, or information. These measures include

- Shifting focus to short-term incentives that may include rebates, regulatory and tax exemptions, or recognition programs for implementing new water-saving technology, and to comprehensive changes that save larger water volumes rather than continuing to provide mainly smaller residential rebates

- Employing regulations to embed conservation into new development and deter excessive discretionary water use
- Creating programs that target users with high potential water savings, and continuing marketing efforts to heighten consumer awareness of water use
- Conducting pilot projects and participating in research projects to identify future conservation strategies and savings potential

Phasing Out Toilet and Clothes Washer Replacement Programs

Toilet Replacement

Beginning in the 1990s, Austin Water offered both rebate and voucher programs to encourage residential, commercial, and multifamily customers to replace existing toilets with water efficient models. With the free toilet program, customers could receive vouchers that allowed them to pick up free high-efficiency toilets from the city's contracted vendor, while the toilet rebate program provided rebates of up to $200 per toilet to assist with the purchase and installation of WaterSense®–labeled toilets. Both programs were very popular and experienced unprecedented participation in their final years, especially in the multifamily sector. The increased multifamily participation was driven by vendors that contracted with properties to facilitate the replacement process.

The Texas Water Development Board's *2004 Water Conservation Best Management Practices Guide* (BMP Guide) encourages utilities to retrofit at least 50 percent of eligible toilets (TWDB 2004). Based on national replacement rate and end use data, combined with program participation, Austin Water estimates that 74 percent of commercial, 93 percent of multifamily, and 79 percent of residential toilets had been replaced by the end of 2011. Additionally, revisions to Austin's plumbing code now require toilets installed in new construction or to replace existing toilets to use no more than 1.28 gallons per flush. As a result, Austin Water ended all toilet replacement programs at the end of FY 2011.

Clothes Washer Rebates

Austin Water, which began offering the residential WashWise Rebate Program in 1998, has modified the program over the years to adjust for new equipment, technologies, and pricing. The most recent program offerings included a $50 water rebate. Over the past decade, water efficiency of clothes washers has steadily increased. In 2004, washers with a water factor (WF) equal to or less than 9.5 were considered efficient. In 2011, changes to Energy Star® criteria required a WF of no more than 6.0 to qualify for Energy Star labeling. These increasingly efficient washing machines are making up a progressively greater percentage of the market. The 2013 revisions to the section of the BMP Guide related to clothes washer incentive programs report that only 4.4 percent of washers sold in Texas had a WF equal to or less than 9.5 in 2002. This BMP Guide shows that by 2009, more than half (56 percent) of all clothes washers available for sale in the United States and 45 percent of those sold in Texas were Energy Star qualified.

According to the BMP Guide, utilities should design residential clothes washer incentive programs so as to increase market share of efficient clothes washers to at least 20 percent by the end of the second year of implementation. The BMP Guide recognizes the increase in efficiency standards and market penetration of these efficient machines that has occurred since 2004. The guide states that due to increased customer acceptance, rebates/incentives may increase the potential for free rider participation in the programs. To avoid continued promotion of an incentive program with high free ridership, the BMP Guide recommends that utilities periodically analyze several factors, including determining local vendor inventories and types of machines for sale; determining local and state market sales; and surveying customers about the types of machines they have. If this analysis reveals that a rebate program has become ineffective, even if the number of rebates issued is high, the BMP Guide endorses adjusting or eliminating the program to ensure proper water savings investment instead of a reward for free ridership.

Water conservation staff conducted a search of area retailers' websites in February and March 2013 to determine efficiency of machines readily available to Austin residents. This search found approximately 80 percent of clothes washers to be water efficient, according to the Consortium for Energy Efficiency (CEE) Clothes Washer Qualifying Product List published in February 2013. A customer survey done by the National Research Center and administered in 2012 also indicated that water efficient washers have become more commonplace. Conducted as part of Austin's participation in the Alliance for Water Efficiency's Residential End Uses of Water Study Update, the survey went out to a random sample of Austin Water residential customers. Of the 118 returned surveys, 65.5 percent of customers reported having a water and energy efficient (Energy Star) clothes washer in their home. Of the 83.6 percent of respondents who reported that their households had taken actions to conserve water in the previous several years, 41 percent reported having installed a water-efficient clothes washer as one of the actions taken. Changes in market conditions that resulted in better availability of water-efficient washing machines and the presence of stronger federal efficiency standards led Austin Water to determine that this rebate was no longer a significant tool for promoting water efficiency. The residential WashWise Rebate ended in June 2013.

THE FUTURE OF AUSTIN'S WATER CONSERVATION EFFORTS

Austin Water is placing increased focus on addressing water use in the commercial sector, reducing water loss throughout the utility system, and lowering discretionary outdoor water uses through customer education and regulatory measures, with a greater emphasis on enforcement.

Incorporating Water Conservation into City Codes

Plumbing Code and Land Development Code Revisions

Austin recently revised its plumbing code and permitting process to make it easier for residential customers wanting to install graywater collection systems to do so. A previous plumbing code revision, effective in October 2010, requires high-efficiency toilets using an average of no more than 1.28 gallons per flush (gpf) and urinals that are either waterless or use no more than 0.5 gpf in all new construction and retrofit projects. Earlier additions to the plumbing code prohibited once-through cooling, commercial garbage grinders, and liquid ring vacuum pumps. In December 2010, Austin City Council approved the Innovative Commercial Landscape Ordinance, which is a change to the Land Development Code that serves as both a water quality and conservation tool. New commercial properties have long been required to install landscaping with the size of the area based on the specifics of the development. Now commercial developments are required to direct a portion of their stormwater to an area that is 50 percent of the required landscaping size. Furthermore, in an effort to limit nonessential irrigation, commercial customers may choose whether to install permanent irrigation in the peripheral regions of the property, and undisturbed vegetation will count toward the 50 percent requirement. Means for conveying stormwater to landscapes vary from passive to active methods, several of which can count toward water quality treatment credit due to their ability to filter water, which can offset the required size of the traditional water quality measure such as a sand filter.

Figure A-24 City of Austin water use compliance patrol vehicle

Water Conservation Code Revisions

The ongoing regional drought created significant challenges for Austin Water and the residents of Austin. Austin Water began a process to revise the Water Conservation Code (chapter 6-4 of City Code), which regulates how customers of the city's water utility may use water in and around their homes and businesses. The goal was to balance conservation of the water supply with the desire to sustain the local economy and the natural surroundings, tree canopy, and vegetation that are unique to Austin. Austin Water encouraged the public, organizations, and businesses to take part in this process by attending workshops to discuss how the city should regulate water use during drought and review specific restrictions, including the watering schedule and limits on car washing. Workshop participants expressed a desire for Austin to implement some conservation measures year-round, begin drought restrictions earlier in a drought, and provide a way for water based businesses to continue to operate during droughts. In 2012 and 2016, Austin City Council approved code revisions, including a permanent year-round, once-a-week watering schedule for all customers, which were developed from comments received through these workshops.

Commercial Facility Assessments

The revised Water Conservation Code includes a requirement that commercial, multifamily, and city municipal facilities equal to or greater than 1.0 acre in size obtain an assessment by an Austin Water authorized irrigation inspector of any permanently installed irrigation system to ensure that the system is not violating the prohibitions against water waste. During an evaluation, the inspector must verify that the irrigation system has no (1) leaks or misting; (2) broken irrigation emitters, spray heads, drip lines, and so on; (3) broken or leaking spray nozzles or bodies; (4) water spray landing on a street or parking lot; or (5) runoff of 50 feet or more or ponding of ¼ inch or more in a street or parking lot. The first group of inspections was due in January 2014. Properties unable to operate their entire irrigation system in compliance with the code by their evaluation due date must submit a compliance plan explaining how compliance will be achieved. Parts of an irrigation system that did not pass inspection cannot operate until appropriate repairs are made.

The amended Water Conservation Code also requires vehicle wash facilities, including self-service car washes, full-service car washes, roll-over/in-bay–style car washes, and facilities managing vehicle fleets or vehicle inventory, that use Austin Water's potable water to have a licensed plumber perform an annual inspection to certify the equipment meets certain efficiency standards. Implementation of this requirement began in January 2014.

Shift to Administrative Enforcement of Watering Violations

In recent years, Austin Water has increased conservation staffing and now has dedicated staff tasked with enforcing the Water Conservation Code on a 24-hour, 7-day-a-week basis. These staff members follow a rotating patrol schedule to look for watering violations that may be occurring. Enforcement efforts are increased in areas with a high number of citizen reports of violations. Figure A-24 pictures one of the city's water use compliance patrol vehicles.

Prior to the 2012 revisions to the Water Conservation Code, watering violations were processed through municipal court as class C misdemeanors. With the revised code, Austin Water sought to change water use violations from a criminal to an administrative matter. Now, customers in violation of the watering restrictions receive fines on their utility bills rather than going through a court process. This shift streamlines the process for fining violators and allows collected fines to remain within the Austin Water operating budget rather than the city's general fund.

Focus on Water Efficiency in the Commercial Sector

Austin Water is expanding its efforts to increase water efficiency among its commercial, industrial and institutional (CII) customers, which represent almost 17,000 accounts and approximately 30 percent of Austin's total water use. Program focus will include learning more about what motivates business customers to implement water saving measures, targeting users with high potential water savings, and continuing marketing efforts to ensure these customers are aware of water efficient technologies and available rebates for implementing them.

Raising Customer Awareness of Conservation Opportunities

Austin Water offers CII customers two opportunities to learn about areas where they can increase water efficiency: the 3C Business Challenge and the Commercial Water Efficiency Audit Rebate pilot program. In the 3C Business Challenge, customers conduct a self-audit of water using equipment. Based on the provided information, conservation staff recommends steps the business could take to reduce water use and provides information about available rebates to assist with water efficient upgrades. The 3C Business Challenge is also a component of the citywide Green Business Leader program initiated by the Office of Sustainability. Customers participating in the Commercial Water Efficiency Audit Rebate pilot program can receive a rebate of up to $5,000 for an independent water audit of their industrial, commercial, or institutional facility. The rebate amount is 75 percent of the cost of the audit, up to $5,000. A rebate is available for each separately metered facility meeting the annual usage threshold of 100,000 gallons per year. To qualify for the rebate, customers must commit to fixing any controllable leaks and making the equipment or system setting adjustments recommended by the auditor. Auditors will also suggest other ways to reduce water consumption and discuss the customer's eligibility for rebates for more efficient equipment and upgrades. This pilot was launched in partial response to a severe, prolonged drought so that leaks could be discovered and fixed and equipment setting adjustments made sooner than would otherwise have happened without the audit.

Expansion of Reclaimed Water System

Since 1974, Austin Water has provided reclaimed water for irrigation and other nonpotable purposes. Using reclaimed water benefits the water system by reducing demand for

Figure A-25 Filling stations make reclaimed water use easier

potable water for nonpotable uses including irrigation, cooling tower makeup, ornamental ponds, manufacturing, and toilet flushing. The city's Water Reclamation Initiative, enacted in 1990, is a plan to expand development of a reclaimed water system to meet current and future nonpotable water demands. In 2009, a drought year with exceptionally high demand, about 5 percent of wastewater received at the city's wastewater treatment plants was treated and reused for nonpotable uses, mostly irrigation of golf courses. In September 2010, the 51st Street Tower, which serves the University of Texas area, was brought online. In November 2011, the reclaimed system was expanded to Austin-Bergstrom International Airport, which is anticipated to save 25 million gallons of drinking water annually. In 2012, reclaimed-water "filling stations" were created so that tanker trucks and street sweepers could use this water for irrigation, road construction, and street cleaning (Figure A-25). The reclaimed system is anticipated to be completed within 25 years with more than 325 customers.

Improved Water Accountability

In 2003, Texas became the first state to require water suppliers to complete a standardized water loss audit every five years. Austin has conducted these audits annually since FY 2007. In 2013, Austin Water reported water loss of approximately 5 billion gallons, resulting in an Infrastructure Leakage Index of 2.72. Most of the loss was the result of system leaks. To reduce water lost to leaks, Austin Water has increased subsurface leak-detection efforts. In FY 2013, Austin Water found and repaired 106 leaks through this leak-detection contract. Austin Water has also begun an accelerated leak response and repair program, which has proven highly successful in reducing repair time. In FY 2006, Austin Water repaired 40 percent of leaks within one day, including 19 percent on the same day. During that year, 55 percent of leaks took more than three days to repair. By FY 2013, response time had improved so that 66 percent of leaks were being repaired within one day, including 32 percent on the same day. Only 15 percent of leaks took more than three days to repair. Austin Water has also worked to improve its water loss audit assessment score, a measure of how valid and accurate the numbers are that the water loss audits calculations are based on. In FY 2013, Austin Water had an assessment score of 77 out of 100 on the AWWA Free Water Audit Software©, version 5.0, with a short-term goal of 80, a score that is sufficient to make informed decisions on what steps to take to further reduce its water loss.

References

Texas Water Development Board (TWDB). 2004. *2004 Water Conservation Best Management Practices Guide.* Austin: TWDB.

Consortium for Energy Efficiency (CEE) Clothes Washer Qualifying Product List. 2013. https://library.cee1.org/content/qualifying-product-lists-residential-clothes-washers

CASE STUDY #11: WAUKESHA, WISCONSIN – COMPREHENSIVE MUNICIPAL WATER CONSERVATION PLAN FOR MID-SIZE SUBURBAN COMMUNITY

Background

Water conservation is an important element in the City of Waukesha long-range water supply strategy of the city of Waukesha, Wisconsin. The historic city (population 70,800) is located in southeastern Wisconsin about 17 miles from Lake Michigan, but about 1.5 miles outside of the Lake Michigan surface water divide. Its current source of water supply is groundwater. For decades, its sole water supply source was the deep confined St. Peter Sandstone aquifer.

Waukesha obtains about 85 percent of its water supply from the deep aquifer. Decades of overpumping by communities in southeastern Wisconsin and northeastern Illinois have created a large cone of depression, with local groundwater levels 500 to more than 600 feet belowground. In addition to greatly depressed groundwater levels, the deep confined aquifer contains radionuclides at concentrations exceeding federal and state drinking water standards. Other contaminants also increase as water levels decline, and long-term use of the deep aquifer is unsustainable.

- To provide water that complies with the radium standards, the city
  - Treats water from the deep aquifer to remove radium.
  - Pumps from a small number of newly developed shallow wells in the Troy Bedrock Valley aquifer. (The shallow aquifer does not contain radium in excess of regulations.) To reduce radium levels, deep aquifer water is blended with water obtained from the shallow aquifers.

This is only a temporary solution because

- Naturally occurring arsenic at levels that exceed the drinking water standard was found in shallow aquifer test wells near the city.
- Pumping the shallow aquifer withdraws groundwater base flows that otherwise would naturally feed local streams, wetlands, and other local environmental features.
- Long-term use of the deep aquifer is unsustainable.

Although conservation may not prevent the need for a new water supply, Waukesha also reduces water demand through a wide range of supply- and demand-side water conservation measures. To conserve existing limited water supplies, the city implemented water conservation measures to conserve water resources and to help meet the goal of providing radium-compliant water at all times. With its initial water conservation plan in 2006, the city set short-, mid-, and long-term conservation goals (see below) and successfully implemented the following:

- Wisconsin's first inclining-block water rate structure
- Wisconsin's first ordinance to ban daytime sprinkling
- Wisconsin's first rebates for high-efficiency toilets
- Information and education campaigns for schools and the general public

However, these efforts alone do not provide a long-term solution to the radium issue. Under a stipulated court order, the city of Waukesha must continuously provide radium-compliant water throughout its public water system by June 30, 2018. Faced with the need for a new sustainable water supply, Waukesha has initiated the process of applying for a Lake Michigan diversion with return flow, in accordance with the Great Lakes–St. Lawrence River Basin Water Resources Compact (the Compact).

The city's conservation plan was updated in 2012 in compliance with the Compact and Wisconsin's new water conservation rule (chapter NR 852, Wis. Admin. Code). A review of the plan's goals, budget, and techniques reveals that the revisions to the city's conservation plan encompass all customer classes and target opportunities for maximizing water savings at the lowest practical cost.

Stakeholders

A diverse group of stakeholders were engaged in the city's 2012 water conservation planning process. Representative business owners, health care executives, developers, landlords, residents, community based organization leaders, educators, and others participated in several stakeholder committee meetings. In addition, the city interviewed some of its largest water users to discuss water use, awareness of conservation measures, and effective ways the city can collaborate with these customers in implementing water efficiency measures.

Conservation Plan Goals

The city's water conservation goals were designed to

- Reduce average-day water demand by 0.5 mgd by the year 2030 and by 1.0 mgd by the year 2050.
- Comply with requirements by chapter NR 852 Wisconsin Administrative Code and by the Compact for implementing environmentally sound and economically feasible conservation measures.
- Align with the Southeastern Wisconsin Regional Planning Commission's long-term water supply plan, published in 2010.
- Incorporate stakeholder and customer input in evaluation of initiatives.
- Monitor performance of water conservation measures to evaluate cost-effectiveness.
- Be inclusive of all city customer classes.
- Target opportunities for highest potential water savings.
- Leverage lessons learned from other conservation programs.
- Promote public awareness of the value of water and need to conserve.

Table A-11 Estimated Costs – Water Conservation Program

Activity Name	2012	2013	2014	2015	2016
Toilet rebates	$5,500	$20,800	$24,900	$33,100	$35,900
Showerhead rebates/installations		$500	$0	$3000	$0
Indoor water use audits	$0	$13,100	$14,400	$14,400	$16,000
Outdoor water use audits	$0	$0	$600	$400	$400
Urinal rebates	$0	$0	$0	$2,900	$3,900
Spray-rinse valve rebates	$0	$2,200	$1,300	$1,300	$2,500
Leak detection mains and hydrants	$7,000	$10,000	$10,000	$10,000	$10,000
Pilot project or tailored incentives	$0	$0	$5,000	$5,000	$5,000
Subtotal	$12,500	$46,600	$51,200	$62,400	$68,700
Public education and outreach	$10,500	$10,500	$10,500	$10,500	$10,500
Program management, auditing, reporting, customer service, sprinkler ordinance	$34,800	$38,000	$45,000	$45,000	$45,000
Estimated program cost total	$57,800	$95,100	$106,700	$117,900	$124,200

Project Implementation Risks

Identified risks in implementing the conservation program were:

- Making additional cash outlays from an already tightly managed budget
- Triggering of an application to increase rates because of reducing revenue attributable to conservation efforts
- Managing perceptions about conservation spending during a troubled economy.
- Estimating savings from fixture replacements when customer discretionary income is low

Recognizing the importance of conservation, the Waukesha Water Utility chose to pursue the program despite the risks.

Budget

Projected program costs for the year 2012 were $57,800, rising to a total of $124,200 by the year 2016. A breakdown of estimated costs is shown in Table A-11.

Partially offsetting these costs, water conservation reduces water and wastewater pumping and treatment costs. These monetary savings are estimated at $700 in 2012, rising to $9,400 in 2016.

Program Techniques Used

Waukesha's water conservation program uses policies, public education, financial incentives, and supply management to promote conservation.

As previously mentioned, policies have been enacted to discourage water waste. Waukesha was the first city in Wisconsin to enact a summer sprinkling ordinance that limits lawn watering, including a ban from 9 a.m. to 5 p.m. Working with the Wisconsin Public Service Commission, the water utility also implemented the state's first inclining-rate structure. The structure encourages households to conserve by charging lower unit rates when smaller volumes are consumed. These rates have been refined and strengthened twice since their inception.

Public education has been a foundation of the Waukesha Water Conservation Plan. The utility promotes its message in schools, at community fairs, in farmer's markets, and by interviewing and publicly recognizing industries that work hard to conserve water.

Arming residents with incentives to reduce consumption and information about the value of water is the focal point of community education programs. The city's most recent initiative, unveiled in summer 2013, involves a web-based program called H2Oscore that allows residents to track their home water use and compare it to neighborhood averages, with incentives for reducing use through discounts at area businesses.

Each year the utility welcomes fifth grade classrooms into one of its well houses to teach them about the water cycle, the threats of pollution, and the importance of conservation. Twenty to 25 sessions are held each year. A working water model developed by students at the University of Wisconsin–Stevens Point provides a visual cross section of the local forces at work on the community's groundwater.

During the US Environmental Protection Agency's 2013 Water Sense® Fix a Leak Week, Waukesha Water Utility supplied teachers with educational packets and surveys. The teachers taught the students to measure the size of their home toilet tanks, check their toilets for leaks, and determine whether their toilet qualified for the utility's $100 toilet replacement program. Every classroom that returned 75 percent of its surveys to the utility received a $100 stipend for classroom supplies. The utility determined that 11 percent of the toilets were leaking and that 22 percent of the toilets were eligible for rebates.

On the supply side, Waukesha Water Utility

- Purchased a leak correlator, trained eight staff members in its use, and held a hydrant survey in 2013
- Performs meter calibration and replacement on a regular schedule
- Manages eight different pressure zones in order to mitigate water and energy loss from unnecessary pressures
- Replaces aging infrastructure at an increasing rate to meet the prescribed rate of 1 percent per year
- Operates the system to regularly achieve only 8 percent water loss

Program Evaluation

As part of the implementation plan, Waukesha has provided for regular and detailed reporting, monitoring, and updating of the conservation program. Among the evaluation strategies included in the plan are database management to facilitate auditing and reporting, regular review of initiatives for effectiveness, annual budgetary reporting of conservation efforts to the Wisconsin Public Service Commission (PSC), and regular meetings to present annual results to the stakeholders.

The city is using 2005 as a base year to evaluate the effectiveness of the program. The city's starting annual volume of total water sales was 2,616,913,000 gallons, while total pumpage was 2,831,510,000 gallons. Average-day pumpage in 2005 was 7.77 mgd, while maximum-day pumpage was 12.87 mgd.

During the first five years of the conservation program, before the 2012 program update, evaluation demonstrated that total water pumped from wells (the only current source of supply for the city) was reduced by 14.0 percent, with peak-season pumping between May 1 and October 1 reduced by 19.4 percent. Declining water use also reduced the number of days in which water demand exceeded 10 mgd from 28 days to 0 days.

Since 2006, the estimated cumulative water savings from the city's conservation and water use efficiency measures is 45.2 MG in 2012. The projected cumulative water savings in 2016 is 89.2 MG.

Residential households that had replaced a toilet in conjunction with Waukesha Water Utility's rebate program during that period are estimated to be saving an average of 9,000 to 11,000 gallons per year, depending on household size.

Using an Alliance for Water Efficiency (AWE) tracking mechanism, the utility quantified the cost benefit of replacing toilet fixtures. Recognizing the impact of this conservation tool, it increased its rebate from $25 per fixture to $100 in 2012.

A financial evaluation of the program in 2012, as presented to the PSC, indicated that conservation program expenses exceeded projections by $25,000. The strong emphasis on education and awareness of the value of water led the utility to invest more heavily than projected in public outreach, particularly in customer outreach and education.

The inclining-block rate structure was implemented for two reasons: to encourage conservation and to communicate the importance of conservation. The efficacy of the blocks with respect to conservation remains to be seen. It may take 10 to 15 years to gather enough data to evaluate a trend. The use of the blocks as a communication tool, however, has been extremely successful. The price of using more water is featured prominently on each of the 80,000 bills sent each year. Furthermore, the inclining rates provide an opportunity for the utility's customer service staff to address conservation during the more than 10,000 calls per year that they receive.

CASE STUDY #12: CITY OF SACRAMENTO, CALIFORNIA – ENGAGING STAKEHOLDERS IN WATER CONSERVATION PLANNING

As of 2015 the City of Sacramento Department of Utilities provided drinking water to approximately 480,000 people, managing 135,000 connections. The distribution system covers a nominal 100 square-mile service area and contains approximately 1,500 miles of pipe. The Sacramento region is expected to grow by about 200,000 people in the next 20 years and the water supply is becoming more vulnerable for several reasons: increasing demand, challenges to the environmental health of the local rivers and the Sacramento–San Joaquin Bay Delta, and escalating costs of delivering a safe and reliable water supply to the community. To ensure that Sacramento remains a good steward of its water resources, the Department of Utilities identified and evaluated strategies to meet the state of California's 2009 mandate to achieve 20 percent gallons per capita per day (gpcd) water use reduction by 2020, and actively sought and considered input from stakeholders on water conservation planning, water efficiency goals, and water conservation measures.

Many experts and interested parties were engaged in developing the city's water conservation plan and met regularly as the Sacramento Water Conservation Advisory Group (SWCAG). The stakeholders were derived from among more than 20 entities that included the Department of Utilities management team; the City Council Water Ad Hoc Committee; and statewide, regional, and community members. In 2012, the SWCAG held 12 meetings and successfully encouraged effective water conservation policy and water use efficiency planning.

The SWCAG was established in November 2010 to serve in an ongoing advisory capacity regarding Sacramento's water conservation programs and policies, and for strategic planning. The Department of Utilities convened with the stakeholders to encourage effective discussion of water conservation policy and water use efficiency, to advance public education and awareness, and to build collaborative partnerships. The California State University Center for Collaborative Policy served as a neutral facilitator. SWCAG members contributed to the city water conservation plan. Specifically, they

- Provided input on water conservation policies and programs in support of the city's water conservation efficiency goals and targets, and 2010–2015 Strategic Plan
- Worked with staff to collaboratively develop a SWCAG "work plan" including key advisory objectives and timelines (2010–2011)
- Provided specific feedback to the Water Conservation Interim Plan (2010–2011) and the Water Conservation Final Plan (2012–2013)
- Assisted in expanding public awareness, education, and technical assistance, and developed and relayed key water conservation messages to the community
- Expanded partnerships with organizations to leverage results

Sacramento's water conservation plan was developed over a two- to three-year period. Significant stakeholder involvement was used to develop the most appropriate plan. The Department of Utilities presented possible goals and measures (a list of 80 water conservation measures) to the SWCAG for comment at various meetings between March and December in 2012. The facilitator assisted the group and encouraged consensus on the goals and the measures. SWCAG and department workgroups were assembled to assist in reviewing, rating, and ranking the measures prior to analysis. The results were provided to the technical consultant, Maddaus Water Management, for the decision support system cost-benefit analyses. (See Tables A-12 and A-13 for details about plan development with stakeholders and stakeholder facilitation budget.)

City staff actively sought and considered input from SWCAG members and other interested stakeholders before finalizing the applicable measures and goals. City staff was open to any and all input from the SWCAG, and retained the flexibility to revise the scope in consideration of new information, advice, or events discussed with the SWCAG. The staff's goal was to direct facilitated discussion toward topics that would significantly benefit the city's water demand management and water conservation efficiency goals within a defined scope and limited resources.

The role of the SWCAG was to provide informed advice to city staff about the city's future water conservation programs and policies as outlined by the water conservation plan. To provide the best possible comment, members learned about existing city policy and practices, water conservation community best practices, and the city's analysis of how best practices might be applied in Sacramento (e.g., the water conservation plan benefit–cost analysis and any constraints in the city of Sacramento context). Members discussed their concerns and suggestions with each other and with city staff, and, where possible, used the facilitator to resolve any differences of opinion within the SWCAG to provide consensus-based recommendations to the city.

SWCAG Member Responsibilities

- Become conversant with the city's water conservation interim plan, goals and best management practice (BMP) targets, mission and vision (2010–2015 strategic plan), programs, and policies (including ordinances).
- Attend all regularly scheduled SWCAG meetings; or, if attendance is not possible, check in with city staff or the facilitator to learn what was missed and notify staff of an alternate who may attend the meetings, keeping in mind that consistent participation is important.
- Review previous meeting summaries and materials distributed in advance of each meeting.

Table A-12 Schedule, City of Sacramento Water Conservation Plan Development With Stakeholders

March 2012	April/May	June	July/August	September	October/December	2013
Meet with director of utilities and division manager to review goals and proposed measures. Bring in neutral facilitation and mediation services	Meet with city manager Meet with department leadership team RFQ for DSS modeling	Meet with City Council Water Ad Hoc Committee	Consultant for DSS modeling hired Meet with City Council Water Ad Hoc Committee	Preliminary analysis of results	Review draft plan with department leadership team	DSS model revisions from stakeholder input Finalize plan Public review Bring to City Council fall 2013
SWCAG meeting agenda: • Set expectations • Process overview – Plan – Analysis process – Timeline • WCP outline • Dept. goals – Review – Feedback • Potential measures – Review – Ask for work group volunteers ◦ Measures ◦ RFQ	SWCAG work groups meet Identify specific measures to discuss: • What metrics should be considered? • What are the issues?	SWCAG meeting Proposed agenda: • Plan update • Work groups report-backs	SWCAG work groups meet	SWCAG meeting Proposed agenda: • Process update • Work groups report-backs • Preliminary analysis of results	SWCAG work groups and implementation work group meet	SWCAG meeting, June 2013

Notes: RFQ = request for qualifications; DSS = decision support system; SWCAG = Sacramento Water Conservation Advisory Group; WCP = Water Conservation Plan

Table A-13 City of Sacramento Stakeholder Facilitation and Mediation Budget, 2012–2013

Facilitator BUDGET

City of Sacramento - Water Conservation Advisory Group - Facilitation and Mediation Services
Feb 1, 2012 - Jan 31, 2013

Labor Category Task Descriptions Labor Resource Billing Rates	Manager Sr Mediator/ Facilitator	Senior Mediator/ Facilitator	Lead Mediator/ Facilitator	Associate Mediator/ Facilitator	Assistant Facilitator	Total Hours by Task	Hrs x rate	Summary
	Staff			Staff	Staff			
	$188	$178	$155	$120	$110			
Task 1: Facilitation & Mediation Services								
	0		96	0	0	96	$14,880	
Subtotal Task 1							$0	
						24	$0	
	0	0	96	0	0			$14,880.00
Task 2: Strategic Consultation								
	0		24	0	0		$3,720	
Subtotal Task 2							$0	
							$0	
	0	0	24	0	0	24		$3,720.00
Task 3: Project Management								
			6	0	0	6		
Subtotal Task 3						0	$930	
							$0	
	0	0	6	0	0	0	$930	$930
Total Professional Services	0	0	126	0	0	126		$19,530.00
Other Direct Costs								
		Qty	UOM trips trips			Unit Rate	Extended Value	
Air Fare			lot			$450.00		
Rental Car						$80.00		
Parking and Mileage						$250.00	$ -	
Subtotal Other Direct Costs								$ -
Administrative Fee (15%) applied to ODCs								$ -
ESTIMATED NOT-TO-EXCEED BUDGET								$19,530.00

- Provide feedback and guidance to staff on the project work plan, upcoming meeting agendas, and other relevant issues.
- Work toward agreement on joint advice where possible by making good-faith efforts to understand differing points of view.
- Provide effective representation by informing his or her organization of SWCAG discussions and consulting them on upcoming issues.

City Staff Responsibilities

The Department of Utilities's Engineering and Operations and Maintenance Divisions provided staff for the SWCAG, and collaboratively created and managed the SWCAG work plan. The city water conservation administrator served as the project manager to ensure timely progress toward meeting the city's water conservation goals. Other responsibilities of city staff were these:

- Accurately convey information about water conservation program and policy issues, as well as resource and time constraints.
- Represent the interest of city ratepayers as well as the Department of Utilities during SWCAG discussions.
- Work with the facilitator to coordinate meetings, including dates and times, room bookings, and audio-visual equipment.
- Prepare informational briefings and materials for SWCAG members.
- Actively seek SWCAG member feedback throughout the planning process and incorporated input.
- Ensure that city policy makers were briefed periodically throughout the planning process and consulted on any pressing issues.

Facilitator Responsibilities

California State University's Center for Collaborative Policy provided a facilitator to

- Provide recommendations to city staff regarding the structure of the SWCAG discussions.
- Actively seek to help all parties express their concerns and recommendations to the SWCAG and city staff, and to help SWCAG members resolve differences of opinion where possible.
- Remain impartial as to the content of the policy and controversial issues under discussion.
- Coordinate with the project manager to develop agendas and meeting materials, and track meeting discussions and outcomes.

Summary of 2012–2013 Stakeholder Process

- Refine plan goals with stakeholders and city staff
- Develop preliminary list of water conservation measures for further evaluation
- Refine water conservation measures with input from stakeholders:
  - Sacramento Water Conservation Advisory Group (SWCAG) and work groups
  - Department leadership team
  - Water Ad Hoc Committee (four City Council members)
  - Decision support system (DSS) model consultant
- Perform benefit–cost analysis of measures
- Select final measures to implement
- Finalize water conservation plan
- Review efficacy of measures to evaluate if they are meeting desired outcome

All decisions regarding city of Sacramento water conservation programs and policies remain the responsibility of the staff as directed by City Council. The SWCAG members were briefed and consulted throughout the city decision-making process and provided input at key decision points in an advisory capacity. SWCAG members did not need to reach consensus on advice to the city; however, the facilitator assisted the SWCAG members in working toward consensus.

CASE STUDY #13: TOWN OF CARY, NORTH CAROLINA – BLOCK LEADER PROGRAM OVERVIEW – ENVIRONMENTAL PROGRAM FOR PUBLIC EDUCATION OF WATER CUSTOMERS

In 1998, the town of Cary, North Carolina, began a Block Leader Program, one of many educational initiatives designed to inform citizens about water conservation. Since the program's inception, more than 330 residents have volunteered to learn about water conservation and distribute information and special promotional items to their neighbors. Block leaders are grassroots environmental educators – resources for their neighbors. They don't enforce ordinances or solicit anything from their neighbors; instead, they keep neighbors well informed and up to date on water conservation issues and practices. Over the years, the program has added information about other town environmental services and programs such as stormwater; fats, oil, and grease; recycling; composting; and sustainability. In 2015 approximately 14.5 percent of Cary's water customers are served by 70 block leaders.

Each year, new and established block leaders attend a two-hour-long orientation session in the spring or early summer, where they learn about the town's summer water conservation campaign, new water conservation initiatives, and the status of relevant utility projects. During the training session, volunteers also pick up materials to distribute to neighbors in their "block." (Block leaders customize their own blocks based on their time availability and level of commitment.) Some volunteers are responsible for the half-dozen homes in their cul-de-sac and others have agreed to be responsible for all the homes in their entire subdivision – for some that can be anywhere from 250 to 700 homes! On average, however, a block leader serves about 80 homes. Staff encourages and appreciates participation at whatever level the volunteer is willing to give.

Volunteers commit to a time obligation of the two-hour orientation plus whatever time it takes to distribute their materials. Throughout the year, town staff maintains regular contact with block leaders through e-mail messages and, more recently, one of the town's Facebook pages. Town news releases pertinent to block leader environmental interests are forwarded to these volunteers, providing them with timely information so they are aware of recent happenings and announcements.

One of the original goals of the town's water conservation program was to reduce residential per capita water consumption by 20 percent by 2015. That goal was achieved in 2012. Because Cary's comprehensive water conservation program incorporates many educational initiatives, financial incentives, and regulations, it is impossible to know exactly how much of the reduction in per capita use is due to the Block Leader Program. However, two studies of this well-established initiative have been conducted as part of Cary's long-term water planning process. In 2007, the Integrated Water Resources Management Plan (CH2M Hill 2007) included an analysis of 2001–2005 water use in two subdivisions, both with block leaders but with differing socioeconomic profiles. Each community was compared to another subdivision with similar economic characteristics but without a block leader. In both cases, water consumption rates were lower for the subdivisions with a block leader. In January 2013, as part of the Long-Range Water Resources Plan (CH2M Hill and Brown and Caldwell 2013), analysis showed that accounts in a block leader area used on average 3 to 33 percent less water than accounts not under the influence of a block leader. Further, during seasonal peak months (April through September) all accounts in block leader areas on average used 3 percent less in 2010 than all other single-family residential accounts in the town's service area; and during seasonal peak months, irrigation accounts in block leader areas on average used 13 percent less in 2010 than all other single-family residential irrigation accounts in the town's service area. As grassroots environmental educators, block leaders are fostering the town's culture of conservation, contributing to the goals of the water conservation program, and making a difference in preserving and protecting our finite natural resources.

References

CH2M HILL. 2007. "Integrated Water Resources Management Plan." Raleigh http://www.townofcary.org/home/showdocument?id=9609

CH2M HILL and Brown and Caldwell. 2013. "Long-Range Water Resources Plan." http://www.townofcary.org/services-publications/plans-publications-reports/long-range-water-resources-plan

Appendix B

G480 WATER CONSERVATION STANDARD CHECKLIST

Utility name: ______________________________

This checklist is intended to help you determine if your utility is meeting the G480 standard. The Requirements column includes a description of the elements of the standard and their relevant section. Review the requirements and provide an explanation in the Documentation column as to how you are meeting the requirement (being specific about source and page numbers may be helpful to you later). Text in **bold** notes verification requirements included in the standard for documents that should be made available to the public to help show how the utility is meeting the standard. Notes on work that still needs to be done before you meet the requirement can be described in the To Do column. Note the date you have met the specific requirement in the Date Completed column.

Section	Requirement	Documentation	To Do	Date Completed
4.1 Regulatory Requirements				
4.1.1	Demonstrate meet or exceed applicable regulatory requirements for jurisdiction	*Note how met*		
4.2 Top Level Organizational Functions				
4.2.1 Staff for conservation initiatives				
4.2.1	Assign dedicated water conservation coordinator	***Provide job description of staff person assigned duties (5.1.1)***		

Section	Requirement	Documentation	To Do	Date Completed
4.2.2 Water conservation planning				
4.2.2	Create, implement, and maintain a water conservation plan	***Provide water conservation plan (5.1.2)***		
	Plan guided by AWWA M52 *Water Conservation Programs: A Planning Manual* or some other guidance	*Note guidance informing plan development*		
	Plan must • Address water conservation across all relevant customer categoriesNote if covered in conservation plan	*Note if covered in conservation plan*		
	Plan should include • Clearly defined and measurable program performance goals • A suite of benchmarks that can be used to assess progress in implementation of the program • A supply assessment • Water conservation strategy • Water conservation goals • Plan evaluation • Ongoing plan maintenanceNote if covered in conservation plan	*Note if covered in conservation plan*		
4.2.3 Water conservation in integrated resources planning				
4.2.3	Treat conservation equally to other water supply options	*Note if covered in conservation plan*		
	Where appropriate, include water made available through conservation as part of the supply portfolio when conducting supply and demand forecasting analyses	*Note if covered in conservation plan.* ***Provide copy of water resource plan that shows how conservation is addressed (5.1.3).***		
4.2.4 Public information and education program				
4.2.4	Develop or incorporate into existing programs information efforts aimed at • Raising awareness • Fostering a culture of conservation and behavior change	*Note if covered in conservation plan*		
	Components of program should include • Effectively communicating the value of water • Information on methods and opportunities for reducing consumption • Delivery of consistent and persistent messages	*Note if covered in conservation plan*		
4.2.5 Water waste ordinance				
	Develop or support creation, implementation, and maintenance of an enforceable water waste ordinance	*Note if covered in conservation plan.* ***Provide copy of ordinance (5.1.4).***		
4.3 Internal Utility Actions and Requirements				
4.3.1 Metering practices				
4.3.1	Implement metering practices that promote conservation, including metering of • All water sources • All service connections	*Note how met*		

Section	Requirement	Documentation	To Do	Date Completed
4.3.1.1 Universal metering	Move toward implementing universal metering of all service (private and public) connections	*Note how met*		
	Establish goal to meter 100 percent of all service connections	*Note how met*		
4.3.1.2 Source water metering	Implement metering of all sources including • Groundwater • Surface water • Reclaimed water	*Note how met*		
4.3.2 Rate structures				
4.3.2	Use a nonpromotional water rate that provides incentive for customers to reduce water use	***Provide document outlining rate structure that shows compliance (5.1.5)***		
4.3.3 Billing practices				
4.3.3	Bill customers based on metered use	*Note how met*		
4.3.3.1 Billing frequency	Bill at least bimonthly	*Note how met*		
4.3.3.2 Reporting consumption	Clearly indicate units for consumption	*Note how met*		
4.3.4 Landscape efficiency program				
4.3.4	Establish a program to improve and maintain water efficient landscapes and irrigation	*Note how met*		
4.3.4.1 Design, installation, and maintenance practices	Develop program intended to maximize water efficiency through proper design, installation, and maintenance of new and existing landscapes and irrigation systems. Programs may include: • Audits • Financial incentives • Design information • Ordinances • Development standards • Education • Examples of how to properly design and operate irrigation systems	*Note how met*		
4.3.4.2 Irrigation scheduling	• Encourage customers to water based upon plant needs • Discourage customers from overwatering or watering during the times of day when water loss to evaporation and wind drift is greatest	*Note how met*		
4.3.4.3 Landscape water budgets	• Where appropriate, implement landscape water budgets to address water use and encourage efficiency	*Note how met*		
4.3.5 Distribution system and pressure management				
4.3.5.1 Water utility audit	Conduct an annual audit of the system using AWWA Water Audit Method, including AWWA Water Audit Reporting Worksheet	*Note if covered in conservation plan.* ***Provide results of AWWA Water Audit worksheet (5.1.6).***		

Section	Requirement	Documentation	To Do	Date Completed
4.3.5.2 Water loss control program	Develop a water loss control program	*Note if covered in conservation plan.* ***Provide written plan for implementation and annual progress report (5.1.7).***		
4.4 External Policy Requirements				
4.4.1 Water efficiency in building codes and standards				
4.4.1	Encourage • Adoption of water efficient codes and standards • Adoption at both state and local level	***Provide evidence that water efficiency is addressed in local building codes for new buildings (5.1.8)***		
4.4.2 Promote water efficient products and services				
4.4.2	Promote the use and maintenance of water efficient: • Products • Practices • Services	*Note how met*		
4.5 Wholesale Requirements				
4.5	Directly implement: • 4.1 regulatory requirements • 4.2.4 public information and education program • 4.3 internal utility actions and requirements	*Note how met*		
	May provide: • Regional coordination on conservation issues and programs • Technical assistance to their retail agencies	*Note how met*		
	May manage conservation activities that are more effectively implemented on a regional scale	*Note how met*		

Appendix C

GUIDANCE FOR MORE IN-DEPTH PUBLIC INVOLVEMENT PROCESS FOR CONSERVATION PLANNING

Identifying opportunities for achieving public involvement includes combining what is known about stakeholder interests and expectations for involvement with what can be done to include the stakeholders in the utility's decision making process. This process is completely scalable based on need for investment in conservation and level of interest in the community to be engaged. Often larger utilities have a more formal conservation planning stakeholder process to elicit feedback and buy-in throughout the planning process. This appendix is focused on a more in-depth public involvement process. Small and medium-size utilities may elect to use elements of the information provided in this appendix. For example, one way to identify strategies for linking stakeholders at smaller utilities to a conservation plan development is to publicize the plan preparation's task schedule. The schedule ideally would show planned public meetings with water supplier advisory committees, commissions, and board or council meetings. These meetings may be the only means to solicit public input in the planning process.

Ideally, initial planning for the more in-depth public involvement begins in the conceptual planning phase of a water conservation program. A realistic assessment of water conservation program alternatives and benefits will be the basis for the planning effort, with conservation being one of the key alternatives. The public's input will have a bearing on the final decision making process. Assuming that the planning process will be collaborative, the development of an effective public involvement program begins with careful, thoughtful planning. Essential elements to successful planning include

- Defining the program goals and assessing potential risks to successful completion of the program (for example, a common risk cited for a conservation program is a reduction in revenue resulting from a reduction in water use; this "perceived" risk is a common misconception that is further addressed later in this appendix
- Assessing the "owner's" (utility or decision making entity) readiness for public involvement
- Understanding stakeholder concerns and expectations for involvement
- Identifying techniques for involving stakeholders
- Developing an optional public involvement work plan

Defining Goals and Expectations of Public Involvement

The term *public involvement* is often used interchangeably with public information, community relations and, in some cases, even marketing. Clearly defining the role the public will play in water conservation planning is more than semantics, and it will help to manage both the public's and the decision makers' expectations.

Marketing is a communication strategy designed to sell the public on a decision that has already been made. Community relations and public information are strategies to distribute information throughout a community, with opportunities for feedback but no real influence over decision making. Truly involving the public in identifying, prioritizing and assessing the value of conservation in the overall water resources portfolio, therefore, will require a process for reaching interested members of the public, educating them on the issues, and providing a meaningful way for the public to express their opinions and influence decision making, enabling them to be effective partners in the decisions that are made.

A necessary step to help manage expectations of the public involvement process (or other communications strategy) is *framing* the project. Determining the specific problem that will be solved and identifying the boundaries of that problem helps to frame the project in terms that both internal and external interests can understand. Facts that establish the need and the scope of the process are clearly defined, and assumptions used in the process are identified. Once these boundaries are understood, it will be easier to establish achievable goals and expectations.

Resist the temptation to develop a solution to a problem too early in the process. It is natural to want to present solutions when describing a problem to the public rather than just communicating about the problem. Instead of bolstering credibility, this position can alienate the public, forcing you to try to sell the project, while pushing the public into positions of support or opposition. At this stage, it is essential that a public forum be established for discussing the problem. In this way, managers can resolve public concerns rather than brokering positions on a project or alternative.

Determine Elements of Process and Opportunities for Public Involvement

For public involvement to be meaningful to both the participants and the decision makers, the public can feed input into the process at logical points. Once the opportunities for public involvement are identified, then planning can take place for appropriate activities or venues.

Stakeholder interviews and open public meetings can help identify community advocates and potential partners to enhance program success.

To gain public confidence, look for opportunities to demonstrate that the water supplier listens and is capable of acting on public input. Start by reviewing with interested members of the public the full range of decisions that must be made during the conservation planning process. The following questions may be helpful in identifying when you would ideally involve the public:

- Do opportunities exist for informing the public about the problem?
- Do the potential water conservation program options require significant public participation to ensure their success?
- Are there opportunities to create other community benefits such as native plant or community gardens or environmental centers related to this program?

Assess Risks and Priority for Each Conservation Program Scenario Being Evaluated

Often, conservation planning is viewed primarily as a low-risk effort simply to use existing resources more wisely. However as water resources become more scarce, water conservation is taking on an ever more important role in meeting future demands. This means that it's important to identify the level of risk. Formally, risk is the potential impact on the utility of the failure of the plan or program. For example, there is significant risk in not developing a plan at all, or not adopting it in time to begin development of resources needed to meet future demands. There can also be potential risks to individual water conservation methods being considered; however, investigating those risks is beyond the level of effort expanded on most plans.

The investment in public involvement may be proportional to the risk posed to an organization that implements a detailed conservation program. Public involvement should not be viewed as an insurance policy against project failure. However, water supply projects that include high risks or significant controversy should include robust conservation program scenarios that have been created with significant public involvement.

In general, conservation programs that use existing water supplies more wisely tend to be viewed as lower risk. For these programs, the public involvement process may be less robust than it would be for more controversial new supply projects.

When necessary, planners assessing a conservation program's risk may consider these variables:

- Consequence of not developing a feasible water conservation plan
- Risks to individual water conservation measures or planning scenarios
- Impact that lack of public involvement may have on internal and external conservation or integrated water resources plan approval

After assessing the risk, focus can be shifted to resources and conservation measures (i.e., mandates or ordinances with costs typically borne by customers) that are more controversial than others and that require public support, and whose success is key to the water conservation plan as a whole.

Determine Public Involvement Priorities and Develop Budget

Public involvement is typically linked with measure implementation, but budgeting for public involvement activities may be aligned with each phase of water conservation planning activities. This will ensure that resources and staff are available to the planning process when needed. Implementation of a public involvement plan is typically "front end loaded," with the bulk of the public involvement activities taking place early in the planning process.

Begin by focusing on the overall plan. Then, consider each conservation program scenario. Any special needs of the individual water conservation program scenarios can be taken into consideration at this time. The risk associated with each scenario will be assessed by simultaneously considering issues (or risk factors) and the priority of the program compared to other scenarios. Issues raised in the past by customers, activists, elected officials, other communities, or the media should be factored in.

Securing the resources necessary to implement meaningful public involvement programs is a fundamental challenge. Because staff and financial resources are limited, the focus is to target future integrated water resources projects with the greatest risk of opposition, delay, or failure. This may or may not include conservation programs viewed by the decision makers and stakeholders on the priority level for investment in public involvement process when compared to vetting more controversial water resources projects.

Assign Roles and Responsibilities

To ensure a credible and sustainable public involvement plan, it is important to assign responsibility for individual public involvement activities to a champion (for example, a public information officer) and/or members of a team. The level of sophistication in the public involvement process will dictate how much time and effort is expended on the public process and how many staff and/or consultants are involved. Determine who has lead responsibility and what support will be needed to provide the appropriate technical input, logistical support, and production assistance for public involvement materials.

A well-planned public involvement initiative includes both labor and expense budgets, perhaps at the task level detail. The number of professional hours for each person and all expenses ideally would be budgeted for each task. Then, evaluate workload and determine whether outside consultants will be needed. They can help level workload without the time or expense of hiring full-time staff, and they offer expertise that may not be available in-house.

A strategic planning process will ensure that conservation alternatives have the appropriate level of public involvement, can help the utility garner internal and external support for public involvement, and will guide the development of a strategic, implementable public involvement work plan, to the extent one is deemed necessary.

Option to Prepare Public Involvement Work Plan

After determining the elements of the planning process that could benefit from interface with public involvement activities, a work plan ideally would be prepared. The work plan could outline specific tasks to be performed in concert with the water conservation and/or integrated resources planning process. These activities include

- Identifying stakeholders
- Identifying public involvement techniques
- Matching public involvement activities with conservation planning elements

Identify Stakeholders

The public is not one group, but many groups or individuals, each with different viewpoints about the water supplier, the problem, and the potential solution. These interested groups or individuals are called stakeholders because they have a stake in the outcome of the decision. It is important to remember that stakeholders are not limited to groups or individuals in favor of the planning process or friendly with the agency. Stakeholders often include opposition groups and vocal detractors. While not always pleasant, including all stakeholders is key to a successful public involvement program. The following questions will help identify key stakeholders:

- Who is likely to be concerned about the problem?
- Who will be impacted by potential solutions?
- Whose support is needed to implement the plan?

Project managers often think they know and understand stakeholder concerns, but these perceptions are sometimes wrong or have changed over time. To ensure that stakeholder views are clearly understood, schedule one-on-one interviews with the people who are likely to shape opinions about the project. Interviews with these opinion leaders often correct assumptions that you might have about public opinion. These interviews are an opportunity to measure the credibility of the organization, further expand the network of project stakeholders, and plan for information dissemination. To conduct a successful interview, come prepared with a list of questions designed to achieve the following objectives:

- Understand a group or individual's specific concerns about a particular problem or set of solutions
- Determine which interests will be satisfied with receiving timely, appropriate information and which interests will seek to influence the decision making process
- Identify other stakeholders to interview (consider sectors of the community that have well defined boundaries or characteristics that may not typically get involved in water resource or conservation planning, but that may have impediments to participation in the final program, such as disadvantaged communities)
- Determine if any of the stakeholder values are contradictory
- Identify where and how interested groups or individuals are likely to find their information

Stakeholders will have varying levels of interest and expectations on any given project. Typically they will fall into one of three groups:

- Group A – expects to be informed but not involved
- Group B – expects to be able to exchange ideas with the water supplier team
- Group C – expects to be able to influence the outcome of the plan or conservation program

Table C-1 Example Worksheet for Stakeholders Interviews

Stakeholder	Issues and Concerns	Expectations for Involvement (A, B, or C)
Resident	Cost and impact on property value	A + B
Local politician	Impact on constituents, tax revenue	B + C
Environmental group	Compliance with regulations, environmental impacts	B + C

Table C-2 Common Water Conservation Plan Evaluation Techniques

Getting Information to the Public	Getting Information From the Public	Exchanging Information with the Public
Newspaper advertisements	Questionnaires or surveys	Information line telephone number
Bill stuffers	Legal notices	Web site
Door hangers	Public comment period	Storefront, trailer, street kiosk
News releases	Focus groups	Talk shows
Op-ed articles and other media partnerships	Informational workshops	Speakers bureau
Utility or community websites	Social media (Facebook, Twitter, LinkedIn)	Door-to-door canvassing
Social media (Twitter, Facebook, LinkedIn)	Web-based comment forms	Advisory groups
Mobile apps		Public meetings/open houses/ workshops
Online exchanges (webinars, Google hangouts, etc.)		Social media (Facebook, Twitter, LinkedIn)
Fact sheets		
Newsletters and blogs		
Displays		
Briefing books		
Digital media: videos, infographics		
Information repositories		
Community events and partnerships with community organizations		
Articles in existing publications		

Understanding the stakeholders and their level of expectations will help in determining which techniques will work best and how to link public involvement activities to project tasks. The example worksheet (Table C-1) will help organize information gained from stakeholder interviews.

Identify Techniques

When selecting public involvement techniques, project needs must be linked with stakeholder expectations, the owner organization's culture, and the time and money available to devote to public involvement. Some techniques are suitable for providing information to the public, others are better suited for two-way communication, while others enable stakeholders to participate in some aspect of project decision making. Water conservation plan evaluation most likely may use some techniques from all three categories. Efforts ideally focus on the most critical tasks associated with the water conservation program alternative scenarios. A few common techniques are listed in Table C-2.

Match Public Involvement Needs and Opportunities with Each Planning Process Element

To enhance support for and implementation of a public involvement process (or work plan), all public involvement activities may be displayed graphically on the project schedule. The goal is to link public involvement tactics with the other project tasks so that public involvement becomes an integral part of the project delivery process.

Understanding the relationship of the conservation program tasks to the concerns of stakeholders is critical to the success of a public involvement program. An effective public involvement program can be crafted once the stakeholder concerns are linked to specific project decisions. The information gathered from stakeholder contact will help link the program tasks with public concerns.

Implement the Public Involvement Plan

Public involvement, like any other project component, requires focused, strategic implementation of a process or detailed work plan. The implementation phase consists of three major steps for the more detailed process and may be streamlined for less involved efforts:

- Chartering the public involvement project team
- Monitoring the project
- Managing change

If the organization already has established project management processes, use them. While it may be necessary to modify the processes to address both the needs of the public involvement program and technical aspects of the project, existing tools are a good starting point.

Formalize a Stakeholder Committee

The first step in implementing the public involvement process and/or work plan is to identify team members and their roles. There may be an internal team and an external committee of stakeholders.

Internal team members might include the program manager, other decision makers (water resource department manager, finance director, or other key managers or staff), project champions, and consultants. Note that senior members of the water conservation planning project team could always be included on the public involvement team. Likewise, the public involvement task leader could be considered a key member of the project management team.

The need for getting the conservation plan adopted may benefit from a new advisory or decision making role for the stakeholders in the form of a committee. These committees or commissions operate best with a charter in place to guide how decisions will be finalized (for example, instruction to review at key milestones and to formally recommend plan adoption to city council or special district board). When chartering an advisory committee, it is helpful to identify an internal project champion. The champion may be a designated senior-level manager who is visible in the community, has access to political decision makers, and has a passion for the planning process to succeed.

Champions have a big-picture view and are able to accommodate temporary setbacks. They consistently communicate both internally and externally about the progress being made, and they are able to maintain project momentum during controversy. External champions are also essential project team members. They are typically allies in the community who act as a conduit about public sentiments. They often serve as contacts and symbols for elected officials, especially during controversial projects, because they demonstrate knowledge of community values, understand project priorities, and illustrate sensitivity to local politics. They can be important local opinion leaders who speak on your behalf.

A chartering session at project startup provides the forum to establish internal team and external stakeholder responsibilities and define schedule, deliverables, budgets, and communication processes. The chartering session is also ideal for developing and endorsing the conservation plan need statement, differentiating facts from assumptions, beginning stakeholder profiles, and if necessary identifying conflict resolution strategies. Ideally, if this conservation planning process is involved in an entire water resources planning project, then the chartering session includes chartering for individual tasks, including public involvement.

Track Progress Through the Process

Predicting and managing process changes involves monitoring the status of a public involvement work plan almost weekly or monthly (or quarterly). Regular communication with project team members facilitates this process. During the monitoring phase of plan implementation, the manager must be aware of the interrelationships between the project's public involvement activities and the schedule of technical tasks to ensure a consistent workflow. Understanding the sequencing of tasks is also important because the plan must build upon successful completion of a preceding task before the next one can be implemented.

Actively manage the schedule and keep team members apprised of upcoming tasks. Active monitoring of the project will indicate where the public involvement program is working and where it needs improvement. Even the most thoughtfully created plan ideally should be assessed periodically to determine how effectively the individual tactics are meeting the intended goals. An assessment focus is on those tactics geared toward addressing the most critical issues or stakeholder groups, since these could presumably stop or significantly delay the planning process. Means of making these assessments include

- Asking key stakeholders their opinions
- Conducting public surveys to measure change in attitude or position
- Tracking media coverage
- Monitoring attendance at public meetings
- Detecting website hits

When developing assessment tools, choose the methods that best meet the needs of the project. Their scientific defensibility may not be as important as their ability to gauge how targeted stakeholders feel. Use the feedback received to correct previous assumptions about opinions on proposed projects.

Manage Change

Change on any project is inevitable. Public involvement programs must address public concerns and perceptions, which are often in flux and can be hard to predict. Even the most subtle change in public sentiment can cause a shift in the public involvement strategy, timing of program elements, and program goals. Public involvement programs must be flexible to respond to events such as election results, pending rules, evolving community issues, and changing stakeholder needs as awareness grows.

Because some changes in public opinion can be anticipated, it may be valuable to assign monitoring responsibility to the most probable risk-related changes. For example, the team member assigned to address political issues should also monitor local elections, candidates, and platforms. Some elected officials are opposed to conservation due to revenue impacts. Ideally, the public involvement team should develop a change management plan that identifies potential changes in project parameters and public opinion, and methods for addressing them. Creating a flowchart that maps out hypothetical scenarios is a helpful way to envision such a plan. Sometimes public information committees have been used to guide needed mid-course corrections so the goals of the public involvement plan are met.

Be prepared for success. Numerous conservation programs have had more rapid customer uptake than originally anticipated. Leaving some contingency funds in the budget can assist managers in rapidly shifting funds to programs that are succeeding without having to go through lengthy budget amendment processes. Also partnerships with stakeholder groups have led to quick uptake of programs, or early lessons learned that required modifications to the original program design. Prepare for flexibility in the plan in order to be prepared to take advantage of these opportunities.

OVERALL BENEFITS OF EFFECTIVE PUBLIC INVOLVEMENT IN CONSERVATION AND INTEGRATED WATER RESOURCES PLANNING PROCESSES

Today's water suppliers must balance multiple, often competing, organizational and political demands while also providing safe, high-quality, and affordable drinking water to their customers. The planning process should seek to frame resource decision making in terms of community values as well as technical and cost criteria. Therefore, the planning process should be designed not only to solicit community input, but also to use that input in a meaningful way during decision making. Benefits of meaningful public involvement include

- Demonstrating the water utility's commitment to efficient investments of ratepayer dollars
- Increasing public access to accurate information about current water quantity, quality, and other affected water properties, thereby increasing trust in the utility and neutralizing misinformation that may be spread through the media or in other ways
- Increasing the public's baseline understanding of the costs associated with treatment and delivery of water, the limits and sources of water supplies, and the need for water reuse and conservation
- Increasing elected officials' comfort level with complex decisions about long-term investments in a community's future by encouraging a public mandate for sustainable, high-quality water
- Reducing the odds that needed permits may be denied and the potential for lawsuits and protracted win/lose battles with project opponents
- Building partnerships with stakeholders to increase reach of programs and achieve greater water savings

Linking water conservation planning and public participation can add credibility to better water resources management. Public involvement can help utilities garner the support necessary to move a water resources project forward. Further, public involvement can help projects succeed by reducing the risk of failure, ensuring compliance with regulatory agencies, and gaining acceptance with the public and elected officials.

This page intentionally blank.

Appendix D

SUGGESTED ELEMENTS FOR DATA COLLECTION TO SUPPORT WATER CONSERVATION PROGRAM MANAGEMENT AND EVALUATION

No.	Data Needed	Comments
1	Total monthly water production	As many years as possible
2	Monthly (potable) total water sales by customer category and the number of accounts billed by meter size	Monthly water consumption by customer class and meter size from as far back as data available to present. (Do not want individual meter records.) Info on billing cycle (monthly/bimonthly). Number of accounts by meter size, class from as far back as possible to present. Use Microsoft Excel format and label categories. Questions to answer: Were there any changes in billing category labels since 2000? When was the billing cycle changed from bimonthly to monthly (since 2000)?
3	List and define the customer categories (all categories)	For example, does single family include duplexes or condos, and what types of accounts are included in each customer category if not obvious? This will be important for the residential per capita water use analysis.
4	Billing system changes	If there were any changes, did you do any retroactive updates to the data?
6	Maximum-day water production and date of maximum day	
7	Annual water production – detailed information on raw and treated water quantity from each supply source	
8	Breakdown of metered customers and unmetered customers by customer class (single family, multifamily, commercial, etc.)	Fully metered – date?
9	Estimates of water loss, as a percentage of total production	
10	Description of any unusual events that may explain trends in consumption data	Such as a large water rate increase or water use restrictions due to a water shortage
11	Water billing rates	As many years as possible
12	The most recent planning documents including general plans and other reports that should be used for water demand analysis purposes.	This list is limited to council/board–approved documents. Only if a plan is currently under development, please indicate if the plan is draft and the proposed adoption date.
13	Nonpotable (reclaimed and raw groundwater) water current use and expansion plans	(1) Total historical monthly nonpotable water use for existing customers. (2) Monthly nonpotable water use data for all reuse water customers. (3) Type of account ("golf course," etc., is fine; we do not need address, account name, etc.).
14	Future job and population situation (for the entire regional service area) – plan to use the water management report with modifications for the current economy	(1) Forecasted growth in population for the service area through 2040. (2) Estimate the number of jobs in the service area for 2011 (and prior years if available) due to abnormal economic conditions. (3) Forecasts of future employment (jobs) in the service area through 2040.
15	Water conservation efforts	(1) Any reports documenting the schedule and scope of implementation and water savings to date. (2) Current status on conversion of accounts to meters and average cost for meter installation. (3) Details on scheduled implementation of future conservation measures (next 5–10 years). (4) Current plans or any considered new development ordinances. (5) Historical annual incentive program and conservation budgets.

No.	Data Needed	Comments
16	Avoided costs	(1) Average annual variable operating costs for water produced or sold (assume none purchased). Costs need to include energy and chemical costs for distributing water and all other variable costs converted to $/million gallons distributed. (2) Cost, capacity, and projected construction dates of any significant planned water capital projects to serve new growth that might be impacted by lowered water demands through water conservation. Information may be generated from the city's current master planning effort.
17	A list of the top 100 customers (accounts) and their monthly and total annual water use for the two to five most recent years(compiled for the region if possible into one file with backup from individual agencies; see comment to the right)	Include type of business, no address needed. Include top 25 from smaller agencies and top 100.
18	Plumbing codes	Schedule of plumbing code changes/enforcement if different than the federal codes. Are there local building codes that supersede the other codes (more efficient)?
19	Distribution network length and breakdown by size and material for each agency, plus number of service connections by agency	
20	Industrial water use changes	
21	Annual population analysis	
22	Weather data	

This page intentionally blank.

Glossary

adaptive plants – non-indigenous plants that easily adapt to the climate and thus require little or no supplemental irrigation once established.

adjusted water budget – an amount of water used to maintain a landscape that is based on area and evapotranspiration rate.

American National Standards Institute (ANSI) – an association involved in developing standards for a wide range of products including plumbing fixtures.

American Society of Mechanical Engineers (ASME) – a professional organization whose Codes and Standards Committee helps to write and publish ANSI standards.

American Water Works Association (AWWA) – a professional organization serving the drinking water supply profession in North America.

American Water Works Association Research Foundation (AwwaRF) – a nonprofit organization that sponsors research for the drinking water supply profession.

apparent loss – water that was not read accurately by a meter, either as unauthorized consumption, including theft, or data analysis errors.

appliance – a water using device such as a washing machine.

audit (end use) – a systematic accounting of water uses conducted to identify opportunities for improved efficiency.

authorized consumption – the annual volume of metered and/or unmetered water taken by registered customers, the water supplier, and others who are authorized to do so.

baseline – an established value or trend used for comparison when conditions are altered.

benefit–cost ratio – benefits and costs measured in terms of money and expressed as a ratio, with benefits divided by costs; typically used as an evaluation tool for different water conservation measures and programs.

best management practice (BMP) – a set of practices, measures, or procedures that are beneficial, empirically proven, cost-effective, and widely accepted by the professional community.

block-rate pricing – a method of charging for water based on the volume used. As more water is used, the price increases (or decreases) through a series of blocks. These pricing structures are designed to encourage efficient use of a resource.

catch-can test – a measurement of precipitation from a sprinkler system in which water is collected in graduated containers (catch cans) placed at evenly spaced intervals for a specific period of time.

central irrigation control – a computerized system that programs sprinkler clocks from a centralized location using a computer.

closed-loop cooling tower – water-conserving cooling tower system in which water used for cooling is recycled through a piping system that cools the water; the water is cooled as air exchanges heat with the pipes.

codes and standards – a set of requirements governing the design and performance of water using fixtures or devices, adopted by a nationally or internationally recognized code setting or standard setting organization.

conservation pricing – water rate structures that increase the price of water as more water is used with the goal of encouraging more efficient use.

consumptive use (evapotranspiration) – combined amounts of water needed for transpiration by vegetation and for evaporation from adjacent soil, snow, or intercepted precipitation. Also called crop requirement, crop irrigation requirement, and consumptive use requirement.

continuous flow system – the continuous use, by an industry, of deionized water to remove contaminants from products and equipment.

cooling tower makeup – water added to the recirculating cooling tower water stream to compensate for water evaporation losses.

cooling water – water typically used to cool heat-generating equipment or to condense gases in a thermodynamic cycle.

cooling water blowdown – procedure used to reduce total dissolved solids by removing a portion of low-quality recirculating water.

cool-season grass – turf grass varieties that are typically not damaged by subfreezing temperatures. Includes bluegrass, Kentucky bluegrass, perennial rye grass, red fescue, and tall fescue.

decreasing block rate – pricing that reflects per-unit costs of production and delivery that go down as customers consume more water.

demand management – the practice of systematically reducing water use for a broad spectrum of utility customers through efficiency measures and conservation, often as an alternative to purchasing new water or expanding water treatment facilities.

demand scheduling – method of irrigation scheduling whereby water is delivered to users as needed and which may vary in flow rate, frequency, and duration. Considered a flexible form of scheduling.

device – a physical item of hardware, such as a new toilet, or specific action by individuals, such as commercial audits, that would save water if the recommendations are implemented or carried out by the water utility or some other group.

discount rate – a value used to annualize costs or bring a future cost to today's cost, often defined to be the cost of borrowed money minus the inflation rate, expressed as a percent.

distribution efficiency – measure of the uniformity of irrigation water distribution over a field.

drawdown – the depth (from the top of the well) to the water in a well when the pump is operating. The water level typically drops when the pump is running.

dual-flush toilet – a toilet designed to use a lower volume of water to flush liquid wastes and a higher volume of water to flush solid wastes.

early-closure flapper – a toilet flapper valve that closes sooner than normal to reduce the volume of water flushed.

effective precipitation – the total depth of rainfall minus the volume lost to evaporation and leaching during a specific time period.

efficiency – a level of water use performance that minimizes water waste

established landscape – a landscape that has been in place for an extended period of time where the roots of the plants are well developed.

evapotranspiration (ET) – water lost from the surface of soils and plants through the processes of evaporation and transpiration combined.

evapotranspiration rate – a measure of the amount of water required to maximize plant growth. This measure is calculated from climatic conditions and factors such as temperature, solar radiation, humidity, wind, time of year, precipitation, and so on.

faucet aerator – a device that can be installed in a sink to reduce water flow rate by adding air to the water.

faucet restrictor – a device inserted into a faucet that forces water through a smaller orifice for the purpose of reducing the flow rate.

fixture – a water using piece of equipment that uses or regulates the flow of water, such as a faucet.

flood irrigation – a method of irrigating in which water is applied from field ditches onto land that has no guide preparation such as furrows, borders, or corrugations.

flow restrictor – a washer-like disk that fits inside a faucet or showerhead and reduces the water flow rate.

flushometer toilet – a tankless toilet with the flush valve attached to a pressurized water supply pipe. These toilets are typically found in large institutional and commercial buildings such as schools, airports, office buildings, and so on.

gauging station – specific location on a stream where systematic observations of hydrologic data are obtained through mechanical or electrical means.

gravity flow – a water system that relies on gravity to provide the pressure required to deliver the water. Consists of a water source located at a higher elevation than the water delivery points.

gravity-flush toilet – the standard tank style of toilet that uses water (at standard gravitational pressure) to perform flushing functions.

graywater – domestic wastewater composed of wash water from kitchen sinks, bathroom sinks and tubs, clothes washers, and laundry tubs that can be used for nonpotable purposes such as irrigation.

green building – a new or retrofitted building meeting certain efficiency standards with respect to water and energy.

green industry – the industry that includes design, maintenance, installation, and management of landscapes.

groundwater mining (overdraft) – pumping of groundwater for irrigation or other uses at rates faster than the rate at which the groundwater is being recharged.

high-efficiency device – a device that achieves a specific function at the lowest water use possible.

high-efficiency toilet – a toilet that uses 4.8 liters of water per flush or less.

high water use landscape – a landscape made up of plants, turf, and features that requires 50 to 80 percent of the reference evapotranspiration to maintain optimal appearance.

increasing block rate – pricing that reduces water use by structuring water rates to increase per-unit charges as the amount used increases.

incremental cost – the additional cost associated with adding a specific amount (increment) of capacity to a water supply.

infrastructure leakage index (ILI) – a dimensionless ratio of the estimated amount of non-revenue water in a system divided by the theoretical minimum for that system (calculated from formulas)

integrated resource planning – a planning process emphasizing least-cost principles and balanced consideration of supply and demand management options for meeting water needs.

irrigation audit – an on-site evaluation of an irrigation system to assess its water use efficiency as measured by distribution uniformity, irrigation schedule, and other factors.

irrigation districts – special units of local government that control the bulk of surface water supplies in the western United States.

irrigation efficiency – the ratio of the average depth of irrigation water that is beneficially used to the average depth of irrigation water applied, expressed as a percent. Beneficial uses include satisfying the soil water deficit and any leaching requirement to remove salts from the root zone.

irrigation plan – a 2-D drawing/plan that illustrates the layout of an irrigation system.

isolation valve – a valve used for isolating all or part of an irrigation system for repairs, maintenance, or wet season shutdown. Common types of isolation valves are the ball valve, butterfly valve, and gate valve.

kilowatt-hour – 1,000 watts of power for one hour, a measure of energy.

landscape area – the total area on a property that contains landscaping elements. Usually equivalent to the total area minus the building footprint and paved driveways and paths.

landscape water requirement – a measure of the supplemental water required to maintain the optimum health and appearance of landscape plants and features.

lifeline rate – a minimum, sometimes subsidized water rate created to help meet basic human needs.

low-flow faucet – a faucet that uses no more than 1.5 gallons per minute at 80 pounds of pressure per square inch.

low-flow plumbing – plumbing equipment that uses less water than was considered standard according to the United States Energy Policy Act of 1992. Must be only type of plumbing sold after January 1, 1994, as amended in 2006.

low-flow showerhead – A showerhead that requires 2.0 gallons of water per minute or less.

low-flush toilet – a toilet that requires 1.28 gallons of water per flush or less.

low-volume urinal – a urinal that uses no more than 0.5 gallons per flush.

low water use landscape – use of plants that are appropriate to an area's climate and growing conditions.

low water use plants – plants that require less than 30 percent of reference evapotranspiration to maintain optimum health and appearance.

marginal-cost pricing – a rate design method in which prices reflect the costs associated with producing the next increment of supply.

matched precipitation rate – an equal rate of water delivery from sprinkler heads with varying arc patterns within an irrigation circuit. Matched precipitation rates are central to achieving uniform distribution of irrigation water.

measure – A device plus a distribution method and possibly an incentive, such as a rebate, targeted at a particular type of end user that, when implemented, will save water.

megajoule – one million joules, a measure of unit energy.

metering – use of metering equipment that can provide essential data for charging fees based on actual customer use.

net present value – the present value of benefits minus the present value of costs.

nonconsumptive water use – water withdrawn for use but not consumed and thus returned to the source.

nonresidential water use – water use by industrial, commercial, institutional, public, and agricultural users.

non-revenue water – water that the utility does not receive compensation for and may consist of the unbilled categories of authorized consumption, apparent loss, and real loss. Also, this value can be derived by calculating the difference between system input volume and billed authorized consumption.

peak-day ratio – a dimensionless ratio of the water produced and distributed on the highest day of the year divided by the average daily value for the same parameter.

peak/off-peak rates – rates charged in accordance with the most and least popular hours of water use during the day.

per capita (person) use – the amount of water used by one person during one 24-hour period. Typically expressed as gallons per capita per day (gpcd).

permanent wilting point – soil water content below which plants cannot readily obtain water and permanently wilt. Sometimes called permanent wilting percentage.

plan – a document that describes a recommended course of action for implementing a conservation program. Usually includes an estimated budget, schedule, and staffing plan.

precipitation rate – the amount of water applied by a sprinkler system in a specific unit of time.

present value – the cost of a future expenditure discounted to today's costs.

pressure loss – loss in water pressure caused by friction of water against the inner walls of pipe or system components.

pressure reducer – a component designed to reduce water pressure in supply system pipe or irrigation lines.

pressurized-tank toilet – a toilet that flushes by using pressure from the waterline entering a pressurized plastic vessel inside the tank.

price elasticity of demand – a measure of the responsiveness of customer water use to changes in the price of water; measured by the percentage change in price.

pricing/rate structure – system used by water utility managers to charge customers for water usage.

pricing signals – rate structures that encourage water conservation by customers.

program – a set of conservation measures planned to be implemented together.

rain sensor – a device that automatically shuts off an irrigation system after a set amount of precipitation falls.

real losses – the annual volumes lost through all types of leaks, breaks, and overflows of mains, distribution reservoirs, and service connections up to the point of customer metering.

recharge – the addition of water to the groundwater supply by natural or artificial means.

recirculating cooling water – recycling cooling water to greatly reduce water use by using the same water to perform several cooling operations.

reclaimed (recycled) water – wastewater that is treated and reused to supplement water supplies.

reference evapotranspiration – the evapotranspiration of a broad expanse of adequately watered cool-season grass four to six inches in height. A standard measurement for determining maximum water allowances for plants so that regional differences in climate can be accommodated.

***Residential End Uses of Water Study* (REUWS)** – the *Residential End Uses of Water Study* published by the Water Research Foundation in 2016

retrofit – replacement of existing equipment with equipment that uses less water.

return flow – that portion of the water diverted from a stream that finds its way back to the stream channel, either as surface or underground flow.

revenue water – those components of system input volume that are billed and produce revenue.

reverse osmosis – common process used to produce deionized water from municipal water.

seasonal rate structure – water rate structure that bills all water consumed during the dry season or peak season at a higher rate than during the other seasons.

self-closing faucet – a faucet that automatically shuts off the water flow after a designated amount of time, usually a few seconds.

simple payback period – the length of time over which the cost savings associated with a conservation measure must accrue to equal the cost of implementing the measure.

simple water budget – a water budget that is the product of reference evapotranspiration, irrigated area, and a conversion factor.

softening – the removal of calcium and magnesium ions from water.

soil moisture sensor – a device placed in the ground at the plant root zone depth to measure the amount of water in the soil. Soil moisture sensors are also used to control irrigation and signal whether watering is required or not.

source protection – protection of a water source, ranging from simple sanitary surveys of a watershed to the development and implementation of complex land use controls, in an effort to avoid water contamination.

spray head – a sprinkler nozzle that delivers water in a fixed spray pattern.

sprinkler heads – devices that distribute water over a given area for irrigation (or to put out fires). The primary purpose of sprinklers, however, is to get golfers wet on cold mornings.

standard – a performance specification for a water using device or fixture.

static water pressure – water pressure as measured when the water is not moving. (The "not moving" part is critical; if the water is moving it isn't static. When measuring static water pressure, all the water outlets on the pipe must be closed. When measuring the static pressure at a house, connect the pressure gauge and then take the reading while all faucets, the ice maker, etc., are turned off.)

submetering – use of separate meters to indicate individual water use in apartments, condominiums, and trailer homes, while the entire complex of units continues to be metered by the main supplier.

subsurface irrigation – applying irrigation water below the ground surface either by raising the water table within or near the root zone, or by use of a buried perforated or porous pipe system that discharges directly into the root zone.

supplemental irrigation – the application of water to a landscape to supplement natural precipitation.

surface water supply – water supplied from a stream, lake, or reservoir.

system input volume – the annual volume input to the water supply system. This equals the volume from own sources plus the water imported volume.

tiered pricing – increasing block-rate pricing.

time-of-day pricing – pricing that charges users relatively higher prices during utilities' peak-use periods.

toilet dam – a flexible rectangular device placed across the bottom of a toilet tank to reduce the amount of water used per flush.

toilet displacement device – a toilet retrofit device (such as a dam, bag, bottle, or rock) used to displace water in the toilet tank in order to reduce the volume required for flushing.

toilet flapper – the valve that controls flushing in a gravity-flush toilet.

under-irrigation – the difference between the water stored in a plant root zone during irrigation and the amount needed to refill the root zone to field capacity.

uniform rate – a pricing structure in which the price per unit of water is constant, regardless of the amount use.

United States Safe Drinking Water Act (SDWA) – federal drinking water quality legislation administered by the US Environmental Protection Agency (www.epa.gov).

variable charge – the portion of a water bill that varies with water use; also known as a commodity charge.

volume from own sources – the volume of water withdrawn (abstracted) from water resources (rivers, lakes, streams, wells, etc.) controlled by the water utility, and then treated for potable water distribution.

water audit/survey – an on-site survey and assessment of water using hardware, fixtures, equipment, landscaping, irrigation systems, and management practices to determine the efficiency of water use and to develop recommendations for improving water use efficiency.

water budget – the amount of water required to maintain plants in a landscape; a method of establishing water efficiency standards by prescribing limits on water applications to landscapes.

water conservation incentive – an effort designed to promote customer awareness about reducing water use and motivate customers to adopt specific conservation measures.

water demand – water requirements for a particular purpose, as for irrigation, drinking, toilet flushing, bathing, clothes washing, and so on.

water efficiency – accomplishment of a function, task, process, or result with the minimal amount of water feasible; an indicator of the relationship between the amount of water required for a particular purpose and the quantity of water used or delivered.

water efficiency measure – a specific tool or practice that results in more efficient water use and thus reduces water demand.

water efficiency standard – criterion creating maximum or acceptable levels of water use.

water efficient landscape – a landscape that minimizes water demand through design, installation, and management.

water exported – the water exported volume is the bulk water conveyed and sold by the water utility to neighboring water systems that exist outside of their service area.

water feature – a pool, fountain, water sculpture, waterfall, or other decorative element that includes water. Many water features recycle water, thus reducing consumption.

water imported – the bulk water purchased to become part of the water supplied volume. typically, this is water purchased from a neighboring water utility or regional water authority.

water losses – the difference between system input volume and authorized consumption, consisting of apparent losses plus real losses.

water right – under the riparian system, a legally protected claim to take possession of water occurring in a natural waterway and to divert that water for beneficial use; under the prior appropriation system, a property or legal claim to withdraw a specified amount of water in a specified time frame for beneficial use.

water supplied – the annual volume of treated water delivered to the retail water distribution system. This equals system input volume minus the water exported volume.

water table – in an unconfined aquifer, the top of the saturated zone; the level at which a well penetrates the top of an unconfined aquifer.

water transfers – selling or exchanging water or water rights among individuals or agencies.

water use efficiency – employing water-saving practices to reduce costs and to slow the depletion of the water supply to ensure future water availability.

weather adjusting irrigation controller – an electronic device that adjusts irrigation station run times and watering days with changes in the weather.

wetlands – lands including swamps, marshes, bogs, and similar areas such as wet meadows, river overflows, mudflats, and natural ponds. An area characterized by periodic inundation or saturation, hydric soils, and vegetation adapted for life in saturated soil conditions.

Xeriscape™ – a trademarked term denoting landscaping that involves the selection, placement, and care of low water use and native ground cover, turf, plants, shrubs, and trees. Xeriscape is based on seven principles: proper planning and design, soil analysis and improvement, practical turf areas, appropriate plant selection, efficient irrigation, mulching, and appropriate maintenance.

Index

NOTE: *f.* indicates figure; *t.* indicates table

L

M

N

P

Q

R

AWWA Manuals

M1, *Water Rates, Fees, and Charges,* #30001
M2, *Instrumentation and Control,* #30002
M3, *Safety Management for Water Utilities,* #30003
M4, *Water Fluoridation Principles and Practices,* #30004
M5, *Water Utility Management,* #30005
M6, *Water Meters—Selection, Installation, Testing, and Maintenance,* #30006
M7, *Problem Organisms in Water: Identification and Treatment,* #30007
M9, *Concrete Pressure Pipe,* #30009
M11, *Steel Pipe—Design and Installation,* #30011
M12, *Simplified Procedures for Water Examination,* #30012
M14, *Backflow Prevention and Cross-Connection Control: Recommended Practices,* #30014
M17, *Installation, Field Testing, and Maintenance of Fire Hydrants,* #30017
M19, *Emergency Planning for Water Utilities,* #30019
M20, *Water Chlorination/Chloramination Practices and Principles,* #30020
M21, *Groundwater,* #30021
M22, *Sizing Water Service Lines and Meters,* #30022
M23, *PVC Pipe—Design and Installation,* #30023
M24, *Planning for the Distribution of Reclaimed Water,* #30024
M25, *Flexible-Membrane Covers and Linings for Potable-Water Reservoirs,* #30025
M27, *External Corrosion Control for Infrastructure Sustainability,* #30027
M28, *Rehabilitation of Water Mains,* #30028
M29, *Water Utility Capital Financing,* #30029
M30, *Precoat Filtration,* #30030
M31, *Distribution System Requirements for Fire Protection,* #30031
M32, *Computer Modeling of Water Distribution Systems,* #30032
M33, *Flowmeters in Water Supply,* #30033
M36, *Water Audits and Loss Control Programs,* #30036
M37, *Operational Control of Coagulation and Filtration Processes,* #30037
M38, *Electrodialysis and Electrodialysis Reversal,* #30038
M41, *Ductile-Iron Pipe and Fittings,* #30041
M42, *Steel Water-Storage Tanks,* #30042
M44, *Distribution Valves: Selection, Installation, Field Testing, and Maintenance,* #30044
M45, *Fiberglass Pipe Design,* #30045
M46, *Reverse Osmosis and Nanofiltration,* #30046
M47, *Capital Project Delivery,* #30047
M48, *Waterborne Pathogens,* #30048
M49, *Quarter-Turn Valves: Torque, Head Loss, and Cavitation Analysis,* #30049
M50, *Water Resources Planning,* #30050
M51, *Air-Release, Air/Vacuum, and Combination Air Valves,* #30051
M52, *Water Conservation Programs—A Planning Manual,* #30052
M53, *Microfiltration and Ultrafiltration Membranes for Drinking Water,* #30053
M54, *Developing Rates for Small Systems,* #30054
M55, *PE Pipe—Design and Installation,* #30055
M56, *Nitrification Prevention and Control in Drinking Water,* #30056
M57, *Algae: Source to Treatment,* #30057
M58, *Internal Corrosion Control in Water Distribution Systems,* #30058
M60, *Drought Preparedness and Response,* #30060
M61, *Desalination of Seawater,* #30061
M63, *Aquifer Storage and Recovery,* #30063
M65, *On-Site Generation of Hypochlorite,* #30065
M66, *Cylinder and Vane Actuators and Controls—Design and Installation,* #30066
M68, *Water Quality in Distribution Systems,* #30068

www.ingramcontent.com/pod-product-compliance
Lightning Source LLC
Chambersburg PA
CBHW081125180926
51611CB00004B/56

* 9 7 8 1 6 2 5 7 6 2 1 3 9 *